D0836573

Christine McClymont
James Barry
Berenice Wood

Published in 1994 simultaneously by:

Nelson Canada,	*and*	The Wright Group
A Division of Thomson		19201 – 120th Avenue NE
Canada Limited		Bothell, Washington
1120 Birchmount Road		98011-9512
Scarborough, Ontario		U.S.A.
M1K 5G4		
Canada		

ISBN 0-17-604368-3   ISBN 0-17-604401-9

1 2 3 4 5 /WC/ 98 97 96 95 94   1 2 3 4 5 /WC/ 98 97 96 95 94

I(T)P ™
International Thomson Publishing
The trademark ITP is used under license

© Nelson Canada,
A Division of Thomson Canada Limited, 1994

Project Manager/Development: Lana Kong
Assistant Editor: Carolyn Madonia
Supervising Editor: Sandra Manley
Art Direction: Liz Nyman
Cover Design: Julia Hall
Cover Photos: ©Michel Tcherevkoff/ The Image Bank Canada
               and E. Honowitz/ Tony Stone Images

Printed and bound in Canada

Canadian Cataloguing in Publication Data

Main entry under title:
Techno-talk

(Nelson mini-anthologies)
ISBN 0-17-604368-3

1. Readers (Secondary). 2. Readers – Technology.
3. Technology – Literary collection. I. McClymont,
Christine. II. Barry, James, 1939– . III. Wood, Berenice L.
(Berenice Laura), 1947– .

PE1121.T43   1994   428.6   C94-931039-5

Series Review Panel

Table of Contents

4

CHANGING THE WORLD

5

ROLL ON, ROBOTS!

Prologue:

THINGS TO COME
Experts Gaze into the Future

by Curtis Slepian

The technological changes predicted in this article may take place in a few years or decades. And some may never happen. After all, no one really knows what the future holds!

Marvin Cetron has an unusual job. He is paid to predict the future.

Cetron isn't some phony fortune teller. He is a futurist—an expert who uses current information to figure out where the country and world are going, and what the future holds.

Technology is changing the world so quickly, we can't keep up. People in fields like transportation and medicine want to know about current advances *and* what advances the future may hold. That way, they can plan for tomorrow—today.

So companies are hiring futurists like Cetron to predict trends that will one day affect their products and businesses. Cetron says, "I look at technology, economics, politics, and social situations and ask, 'What will the future probably look like?' "

Cetron doesn't take wild guesses. First, he gathers thousands of statistics (figures) and feeds them into big computers. Then, with the help of experts, he studies the computer readouts and makes his forecasts.

Predicting things like weather trends and population growth is pretty scientific. That's because these forecasts use huge amounts of numbers and calculations. But, according to Timothy Willard, forecasts about society are less scientific. You see, people are a lot more, well, unpredictable.

Willard is the former editor of a magazine called *The Futurist*. In it, experts write about future trends in all areas—from computers and technology to lifestyles and education.

Sometimes predicting the future is pretty tough. For instance, Willard says, "In the 1970s, futurists predicted marriage would soon end—and that hasn't been the case at all."

Thinking about the future is good sense. "In order to have the future you want, you must figure out what you want and then help create it," says Andrew Lawler, another former editor at *The Futurist*.

Adds futurist Gareth Branwyn, "Studying the future and deciding what future we want and working to make those changes—that's the thing that's going to help us."

Here are some peeks at the future, from *The Futurist* magazine. Some things may take place in a few decades or less. And some may never happen. After all, no one really knows what the future holds!

Tiny robots may perform surgery inside a patient's body. After a patient swallows the "microrobot," a human surgeon will guide it to the trouble spot. The doctor will guide it with the help of a 3-D computer simulation of the patient's insides.

*

Santa won't be the only person with a North Pole address. In the future, people will be moving to the world's polar areas, because new technologies will make it easier to live in snow, ice, and low temperatures.

*

Future astronauts won't be sipping bottled mineral water—they'll be drinking asteroid water! Because water is too heavy to ship into space in large amounts, other sources for it must be found—like asteroids! Asteroids have no atmosphere, but they do have clay-based materials that hold water and may be mined for drinking water.

*

Tomorrow's hot fashions will be designed with the environment in mind. For example, awesome future clothes may be head and shoulder coverings that shield wearers from the sun's dangerous ultra-violet rays. And people might walk down the street proudly wearing a transparent helmet that keeps out smog and bad odours.

*

One day, the blind may see. Scientists hope to give sight back to the sightless by transplanting light-sensitive cells into their eyes. Already, researchers have done similar transplants in the eyes of rats.

*

Computers may one day be living things. Proteins are the building blocks of living tissues. And scientists have made protein molecules that can conduct electricity. Eventually, they may be placed inside computers. These "living biochips" will be smaller, faster, and less hot than silicon chips. Special living biochips could sense sound or touch or smell. For example, one of these biochips could be able to "smell" poison gases in a factory and shut down the equipment.

*

Hypersonic trains will take people from New York to Los Angeles in half an hour. They might even travel across the Atlantic in an hour! How? The "trains" would travel along a tube. When air is pumped out of the tube, the trains would "fly" through the vacuum that was created.

*

Real food will be produced artificially. Instead of growing whole plants in the ground, certain cells of the plants will be grown separately. These cells will produce only a certain part of the plant, like the inside of an orange. This way, you can get orange juice without the skin—and without trees!

*

In 30 years, it's possible that a space hotel will be open for business. Tourists will be able to take space walks and play zero-gravity sports, as well as go on side trips to the moon. Guest rooms will have artificial gravity, so taking a shower won't be a washout.

*

The real danger in the future may not be drugs, but "virtual reality." Virtual reality is a 3-D world created by a computer. People can enter and control this "world" by wearing goggles and a special sensor glove. Virtual reality already exists. But when the computers that create it become more powerful, people might not want to leave the simulation. So staying in it too long might become a crime!

1

Spinoffs from Space

▲▽▼▽▲▼▽▼▷▲▼▷▼▲▼▽▼▲▼▷▼▷

Man on Moon

Stanley Cook

▲▼▷▼▲▼▷▼▲▼▷▼▷▲▼▷▼▲▼▷▷

▲ ▼ ▶ ▼ ▶ ▲ ▼ ▶ ▼ ▶ ▲ ▼ ▶ ▼ ▶ ▲ ▼ ▶ ▼ ▶ ▶

SPINOFFS FROM SPACE

by Patricia Barnes-Svarney

New devices developed for the space program have had some surprising spinoffs
down here on Earth.

How far can you reach? The Canadarm, also
known as the Remote Manipulator System at
NASA, is the 15-m extension arm you have seen
space-shuttle astronauts use to launch or retrieve a
satellite. Developed by Canadian scientists, the
Canadarm took its first trip into space in 1981
aboard the space shuttle *Columbia*. Did you know
that this type of arm, based on space-shuttle
design, also is used to mine rock and to travel into
dangerous parts of a nuclear reactor?

The breathing apparatus worn by firefighters
to protect themselves against smoke inhalation and
injury is adapted from the life-support systems
used by the Apollo astronauts on the moon. Even

the lightweight material, the face masks, and the warming device on the air bottle came from Apollo and other space-suit designs.

Do you have problems with fog on your goggles when you ski? Try using electrically heated ski goggles, a spinoff from the space program. Thanks to the heated visors used by the astronauts, moisture does not condense, so the visors provide fog-free sight during any activity—and you can see going down a snowy hill!

What kind of paint will withstand heat from the launch of the *Saturn V* or the space shuttle? K-Zinc 531—a type of paint that protects not only the launch gantries[1], but also protects the Statue of Liberty from corrosion.

▲ ▼ ▶ ▼ ▲ ▼ ▶ ▼ ▲ ▼ ▶ ▼ ▶ ▲ ▼ ▶ ▶

SPACE Q AND A

from Odyssey magazine

I heard that if you get close to a black hole, you will get pulled in and never come out. Is this true?

My old friend Isaac Newton showed that if you go from where you are to halfway toward any place with gravity (a planet, star, or black hole), that object's gravitational force increases by four times. So you will need four times as much force in your rockets to get away. Move even closer, and you will need even more force to get away. Around a black hole, there is a place called the *event horizon*, where the gravity is so strong no amount of rocket power will get you free. If you get that close to a black hole, you will never get out to tell what you saw.

When will all the planets be in alignment? Will it happen in our lifetime?

Not once in recorded history have the planets lined up, all nine in a row, like a flock of ducks. Planets align all the time, though usually just three at a time. Often the earth and two planets put themselves in a line, and we get to see a conjunction of planets in our night sky. At these beautiful occasions, you actually can watch one planet move past the other over a period of several days. Then you can really see why the ancient Greeks named these "planetes," the wandering stars.

I wonder if commanders or pilots can fly a shuttle orbiter if they do not have 20/20 vision. Can they wear glasses up in space?

Yes, they can. Rockets push tremendous G forces[2] onto a pilot upon takeoff, making everything weigh two, three, or more times its normal weight—even glasses. But glasses are so light that even three times the weight is no problem, and a pilot is too busy upon launch to care. In a few minutes, he and his glasses will be weightless anyway, and then he will care—to hold his glasses on with a strap.

How do you take a bath in space?

Can't stand to wash, huh? Then the shuttle is the place for you, because you can't take a bath there! Most astronauts are too busy during their short stays in space for one anyway. But if someone feels too gritty, he or she can wet a sponge or towel in an enclosed bubble in which water comes in one side and is vacuumed out the other. Then, curtained off in the lavatory, the astronaut can sponge clean and towel dry. Actually that's a pretty ecological way to bathe down here, too!

1. **gantries:** towerlike structures, with platforms at different levels, used for servicing a rocket on its launching pad

2. **G forces:** a G is a measurement of acceleration. One G equals the acceleration produced by the gravitational force at the earth's surface (9.8 m/s^2).

▲ ▶ ▼ ▶ ▲ ▼ ▶ ▼ ▶ ▲ ▼ ▶ ▼ ▶ ▲ ▼ ▶ ▼ ▲ ▶

SATELLITES IN SPACE

by Jack R. White

Exciting as it is when astronauts blast off into orbit, unmanned satellites and spacecraft actually do most of the work in space.

T minus ten seconds and counting ... we have a go for main engine start ..." The voice of the range control officer at Kennedy Space Center crackles over the speaker in the space shuttle cabin. You scan the switches and computer display on the instrument panel in front of you. Everything looks right. All systems are "go" for launch.

"... seven ... six ... we have main engine start ..." A deep rumble begins to shake the ship as the shuttle main engines fire.

"... two ... one ... we have SRB ignition ..." The two SRBs, solid rocket boosters, fire. The thrust pushes you back into the padded seat.

"... lift-off, we have lift-off ..." The shuttle is on its way into space once again.

Higher and higher and faster and faster you climb. Two minutes after launch, the solid rocket boosters burn out and separate from the shuttle to parachute back to Earth. Five minutes later, the giant external tank strapped under the belly drops away and the main engines shut down. Then there is silence.

Your stomach has the same funny feeling you get coming down in a roller coaster or a fast elevator. A pencil floats up out of your pocket and hangs in mid-air. Now you know you are really in space.

The sky outside the viewing port has changed from blue to inky black. Overhead (because the shuttle normally flies "upside down") is the great shining ball that is the earth. Except for the swirls of white clouds, Earth looks very much like a model globe without the stand. Passing over Africa now, you can see completely across the continent, from the Atlantic Ocean on the western side to the Indian Ocean on the east.

Shades of light greens and dark blues mark shallow and deep areas in the ocean. A mountain range cuts across the landscape like wrinkles in a great green blanket. A whirlpool-shaped cloud over the Indian Ocean indicates a storm gathering strength there before it strikes toward land.

Here and there a few signs of human life can be seen on the planet outside your window. Humans usually make their mark upon the land in straight lines. A thin silver band is a highway. Rectangles of brown and green are farmers' fields. Streaks of white are the vapour trails of jet airliners.

This View of Earth

It is this view from high above the earth that makes space so valuable. From this high position, radio and television signals can be relayed between Earth stations on nearly opposite sides of the world at any time in any kind of weather. Being high enough to see patterns in the clouds that can't be detected from the ground or from airplanes makes it possible to spot storms at sea long before they can reach land. Weather maps and measurements from around the world can be obtained from space in minutes.

Maps of the land can be made from space faster, cheaper, and more accurately than by land surveying or by aircraft photography. Plant life and mineral resources can be mapped and measured even in deep jungle or high mountain areas that are difficult to reach on foot or even by airplane.

Space has another special feature: no air. Air causes problems for astronomers trying to get sharp pictures of planets and stars. Air bends the light rays and smears the images seen through Earth telescopes. A telescope in space can see farther and more clearly than any Earth-based telescope can. One of the newest and most exciting uses of space today is for astronomy.

Early Dreams of Space

Long before rockets with the power to make it into space could be built, science fiction writers foresaw many of today's uses of space. In 1869, a story in the *Atlantic Monthly* by Edward Hale entitled "The Brick Moon" described a space station made of brick to be used to help ships navigate. Since radio was unknown then, people on this space station communicated with Earth in a kind of sign language by jumping up and down in long leaps matching the "dashes" and short leaps matching the "dots" of Morse code. These signals could be seen on the ground by viewers with telescopes.

In 1945, an article by Arthur C. Clarke in *Wireless World*, entitled "Extra-terrestrial Relays," proposed space stations manned with technicians who would receive television signals from Earth and broadcast them around the world. Clarke was one of the first to realize the importance of an orbit 35 880 km high. At this height, a spacecraft circles the earth at exactly the same speed as the earth turns, so it always stays above the same place on the ground.

Although many people guessed how useful space would be, most early plans called for men and women to be in space to operate the equipment. Few people dreamed of the advances in electronics that would make it possible for machines to work in space without human control. Today, unmanned spacecraft (those without people) do most of the work in space. Sending people into space has been too expensive to do very often. For every rocket launch carrying astronauts, there have been hundreds of launches that carried only instruments.

Since the beginning of the space age in 1957, unmanned spacecraft have brought us more information about our earth, our solar system, and our universe than perhaps any other source. Future history books may very well record that the most important invention of this century was the artificial satellite.

Satellites of Earth

What is a satellite? The word "satellite" means any smaller object that orbits (circles) a larger one. The moon is a natural satellite of Earth. A satellite is held captive to its planet by a balance between two forces that pull in opposite directions like an invisible game of tug-of-war.

The planet's gravity pulls the satellite downward. The force created by the speed of the satellite pulls it upward. As long as the speed doesn't change, the forces stay balanced and the satellite will circle in orbit forever. Artificial satellites (those made by humans) and natural satellites both obey exactly the same laws of gravity and motion.

Today, artificial satellites crowd the skies overhead. There have been over 2600 successful space launches over the years. NASA, the National Aeronautics and Space Administration, is now keeping track of over 5000 objects in orbit. Many of these objects could be called space litter. They are bits and pieces of satellites and of rocket boosters that have exploded in orbit.

Most of the satellites in orbit are either small or far from Earth, but about 200 are large enough and low enough to be seen with the eye. Have you ever seen a satellite? You don't have to have a telescope or binoculars. All you need is a clear, dark night and a little patience. Watch the sky an hour or two after sunset and there is a good chance you will see a satellite or two. They look like faint stars moving at high speed across the sky.

Satellites in Our Lives

Did you use a satellite today? Satellites and the information we get from satellites are used by almost all of us every day. If you watched a movie channel on cable television or the news or a sports program from overseas, if you listened to a weather report or saw a weather map in the newspaper, if you made an overseas telephone call (or even some long distance calls within North America), then you probably used a satellite. These are only a few of the ways satellites bring information and entertainment into our lives.

The thousands of satellites come in all sizes, shapes, and uses, but there are six main kinds:

- *Communications satellites* relay television and telephone signals around the world.
- *Weather satellites* send down pictures of cloud patterns to warn of storms and make many other weather measurements.

- *Environmental satellites* make maps of the natural resources of Earth.
- *Astronomical satellites* photograph and make scientific measurements of the sun, the stars, and the other planets for astronomers on Earth.
- *Spy satellites* scan the earth and sky to warn of missile launches and to photograph military bases.
- *Navigation satellites* help ships and planes find their ways safely.

The satellites of today affect our lives directly and indirectly in dozens of ways. The satellites of tomorrow will be used in even more ways. We will have direct links into space from our homes and vehicles for communication and navigation. We may even wear wrist radio communicators the size of a watch that will let us talk to anyone anywhere by satellite. Much of our energy problems may be solved by giant satellites that collect energy from the sun and beam it down to earth as microwaves. In hundreds of other ways, satellites will find uses that are still undreamed-of today.

▲ ▶ ▼ ▲ ▶ ▼ ▲ ▶ ▼ ▲ ▶ ▼ ▲ ▶ ▼ ▲ ▶ ▼ ▶ ▶

WEIGHTLESSNESS TRAINING

by Gloria Skurzynski

How do you train astronauts to feel comfortable and do useful experiments while floating around a space shuttle cabin? By simulating weightlessness here on Earth.

Astronauts' jobs are different from anyone else's, because they don't work on land or sea or in the air. They work in space.

In the almost-zero gravity of outer space, everything flies around if it's not tied down. Cheeks and eyebrows rise higher on faces, spines stretch out a few extra centimetres, and shoulders hunch up because the arms don't weigh them down. Astronauts need to practise being weightless. But how is that possible in Earth's gravitational field?

There are two ways to simulate weightlessness without leaving Earth. One is to enter a neutral buoyancy pool: At NASA's Lyndon B. Johnson Space Center in Houston, Texas, the pool is called WETF (pronounced wet-if), for Weightless

Environmental Training Facility.

Getting ready to go into the WETF is a big job. Several people have to help an astronaut into the bulky space suit. After all the lower layers are on and the space suit is tightly sealed, air is pumped inside to a pressure of about twenty-five kilo-pascals. If the astronauts were lowered into the pool just then, they'd float like inner tubes because the air-pressurized suits are lighter than water. To create neutral buoyancy—weightlessness—lead weights are added to the front and back of the suits, and to the arms and ankles, the way scuba divers wear weights to let them sink.

Next the astronauts are lifted by a hoist and lowered into the 7.6-m-deep pool. Also in the pool are mock-ups of the payload bay or other hardware such as the Hubble space telescope. Underwater, the astronauts practise repairs they might have to make on equipment outside the spacecraft (inside it, they wouldn't need space suits).

Space suits are uncomfortable and tiring, even in zero gravity. Says astronaut Robert Stewart, "If the suit is fit properly, you're kind of wedged into it."

Astronaut Kathryn Sullivan adds, "You learn, through several runs in the water tank, a whole set of lessons that have to do with suddenly being a person of greater mass and volume." This training is essential because, Sullivan says, "The vacuum of outer space is a tough and unforgiving environment." Simulating outer-space conditions in the

WETF lets astronauts practise for emergency situation ahead of time.

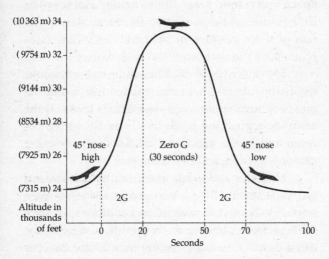

Parabola Flying: During the curve, or parabola, everything inside the KC-135 aircraft is weightless.

The second way to simulate weightlessness is parabola flying (see diagram). NASA's KC-135, a former Air Force refuelling tanker, flies upward at a 45-degree angle to about 30 000 feet [9144 m] altitude. Then it curves around in a parabola and starts to descend at a 45-degree angle. For the 30 seconds or so that the airplane is at the top of the curve, everything inside it is weightless. [The reason? The downward force due to gravity is cancelled by the upward centrifugal force—the same force you feel when you turn a corner in a car.]

In these simulations the astronauts don't need space suits; they wear their regular clothes. The first couple of times people fly parabolas, about half of them get motion sickness: NASA provides white plastic airsick bags, and they're used!

Mostly, KC-135 flights are practice sessions for astronauts who will make space walks. Putting on space suits while weightless can be tricky! If the astronauts push too hard, they may fly across the cabin and bump into someone. This sets the other person spinning, and could cause injury.

During parabola flights, astronauts also try out experiments they'll perform in space, to learn how objects react in weightless conditions. In spite of the serious purpose of the flights, people in the cabin have a lot of fun floating—for 30 seconds per parabola. Then gravity sets in again.

The weightlessness they experience in these training flights isn't simulated: The trainees *really* are weightless during the parabolas. What's simulated is the environment of space. Inside the shuttle, in orbit, a pen accidentally dropped can drift away, and water spilled out of a glass floats in spheres all over the cabin.

▲ ▼ ▶ ▼ ▶ ▲ ▼ ▶ ▼ ▶ ▲ ▼ ▶ ▼ ▶ ▲ ▼ ▶ ▼ ▶ ▼ ▶ ▶

THE SPACE SHUTTLE DISASTER

by James McCarter

It should have been a successful, routine shuttle mission. The *Challenger*'s main objective was the launch of a communication satellite that would provide a continuous link between the shuttle and mission control.

The public, however, was most interested in Christa McAuliffe, a public-school teacher and the first non-astronaut to go into space. At lift-off time, millions of people (including McAuliffe's students) were glued to their television sets.

Shortly before 9 a.m. on Tuesday, January 28, 1986, the crew of the space shuttle *Challenger* took their places for mission 51-L. Twice before, the launch at the Kennedy Space Center in Florida had been postponed due to technical problems. On that Tuesday it had been bitterly cold at dawn, and the crew were preparing to put off the launch again. However, at a weather briefing after breakfast, they were told that lift-off might be possible later on, at around noon.

Alongside the fixed service structure[1] on Pad 39B, the black and white orbiter[2] gleamed in the winter sunlight, attached to its two solid fuel rocket boosters and the huge orange liquid fuel tank. But the mission management team at NASA was worried. Sheets of ice and icicles hung around the launch pad. Would the cold affect the shuttle's complex systems? Could falling ice damage the shuttle at lift-off?

The countdown was stopped at T minus nine minutes (nine minutes before lift-off) while technicians assessed the problem. Inside the orbiter the crew waited for a decision. Five of them were experienced astronauts—Commander Dick Scobee, pilot Mike Smith, and three mission specialists[3], Judy Resnik, Ellison Onizuka, and Ron McNair. The other two crew members were not astronauts, and had been specially selected for this mission. They were Greg Jarvis, an engineer, and Christa McAuliffe, a teacher.

At 11:15 the final decision was announced. There was no danger. The countdown would recommence at 11:29 for lift-off at 11:38.

"Malfunction"

At the Launch Control Center, Cape Kennedy, the friends and family of the crew and the VIPs took up their positions in the grandstand to watch lift-off.

On board *Challenger* the five computers ran through the launch sequence. In firing room three[4] at mission control the launch technicians watched as the read-outs flashed on their VDUs[5]. At T minus six seconds the crew heard the high whine of the main engines starting up, deepening to a fierce roar as they reached full thrust. Then came the blast of the two solid fuel rocket boosters. Amid huge clouds of steam and smoke, *Challenger* rose on a column of fire.

Thirty-five seconds into its flight, the shuttle passed through the sound barrier. At about this point the craft hit violent crosswinds and the computers automatically made continuous adjustments to the flight surfaces, to keep *Challenger* on a true course.

Unknown to the computers and unseen by anyone, a small tongue of orange flame began to flicker around one of the joints on the right-hand booster. It grew, licking one of the struts that connected the booster to the liquid fuel tank. By 72 seconds hydrogen from the now leaking fuel tank ignited.

The booster swivelled on its mooring and pierced the top of the tank containing oxygen. At a height of 13 km above the Atlantic, hundreds of tonnes of liquid fuel ignited into a huge fireball. The orbiter was flung off course, spinning at immense speed, to be torn apart almost immediately by huge forces in the atmosphere.

The crowd on the grandstand were silent.

"Flight controllers here are looking very carefully at the situation," said a voice over the loudspeaker. "Obviously a major malfunction ..."

World in Shock

In the firing room at mission control the explosion suddenly crackled over the communications channel. The TV screens showed a billowing cloud and spinning wheels of fire. Thin trails of steam and smoke made a cobweb in the sky as pieces of wreckage fell to earth.

NASA's horrifying film of the tragedy was soon flashed onto TV screens worldwide. People watched it again and again in shocked disbelief. What could have gone so dreadfully wrong?

What Went Wrong?

Examination of film of the launch taken by cameras on the ground soon made clear what had caused the tragedy. About 58 seconds into the flight the fatal sideways spurt of flame from the right-hand booster was clearly visible. At that point *Challenger* was doomed. Further film of the shuttle immediately after lift-off revealed more evidence—puffs of smoke seen escaping from the joint between the two lower segments of the solid rocket booster's casing.

Engineers at the Thiokol company, the firm which manufactured the boosters, had for months been expressing their concern that the joints between the segments of the casings might fail under certain conditions. Some had even said that the shuttle should not fly until the problem was solved. The Rogers Commission soon began to concentrate its investigations in this area.

The joint between the segments of the booster casing is relatively simple—a protruding metal tang, or rod, slots into a clevis[6] on the lower segment. The joint is sealed with two rubber O-rings, one to act as a back-up should the first O-ring fail. These rings were designed to be forced into place by the pressure of hot gases when the booster is ignited.

However, on more than half of the shuttle missions the O-rings were worn away by the hot gases, in some cases seriously so. Tests had also shown conclusively that the O-rings were more liable to fail in freezing conditions.

The Rogers Commission made its report. What had happened was that the O-rings had failed on ignition of the booster—hence the puffs of smoke seen escaping. The intense heat caused the material of the casing segments to weld together. The weld was airtight, but very brittle. The strong buffeting from the high crosswinds vibrated the shuttle, causing the weld to break at the 58-seconds mark. On a calm day, the shuttle might have survived. But on that cold, windy January day, the failure of the O-rings had disastrous consequences.

The fatal flaws in the design of the segment joints of the booster rockets could be put right. But there were also flaws in the management of the project which allowed the risky launch of *Challenger* 51-L to go ahead at all.

The first lesson to learn from the tragedy is that the best, most reliable technology takes time and a great deal of money to develop. Secondly, the world of space technology is best understood by specialist engineers, rather than politicians or business people. The engineers should always have the final say as to whether a launch should proceed or be cancelled.

1. **fixed service structure:** the massive structure that supports the shuttle on the launch pad. It carries the pipes and cables that allow the shuttle to be fuelled and powered before take-off.

2. **orbiter:** officially, the space shuttle is the term for the entire craft, including the fuel tank and rocket boosters. The orbiter is the airplane part of the shuttle that goes into orbit and returns to Earth.

3. **mission specialists:** the astronauts who conduct experiments and launch satellites, rather than those who fly the shuttle

4. **firing room three:** the operations room at the Launch Control Center, Kennedy Space Station

5. **VDUs:** visual display units (computer screens)

6. **clevis:** a U-shaped piece of metal with a bolt or pin, used to fasten two objects together

▲ ▼ ▶ ▲ ▼ ▲ ▼ ▶ ▲ ▼ ▶ ▲ ▼ ▶ ▲ ▼ ▶ ▲ ▶

LAST RIDE

Andrea Holtslander was a high-school student in 1986 when the *Challenger* space shuttle exploded.

> We watch in horror as
> the booster rockets twist
> crazily through the sky
> like balloons
> whipped free
> from a child's grasp.

The horror is the reality on the screen.

"On this day of tragedy … we watch in horror as …"
> And for the benefit of those who
missed the live show
> > we will run the
> > fireworks once again.

> The spotlight moves to the grief-stricken
families and we can have our
> > > > heart-strings pulled
> > with 20 million others
> as we watch their
> tears fall,
> > > LIVE.

Having wrung all the tears from his
audience the ringmaster can now turn to sports
as seven families try to put
together their lives
scattered over the Atlantic
Ocean.

Andrea Holtslander

▲ ▶ ▼ ▲ ▶ ▼ ▲ ▶ ▼ ▲ ▶ ▼ ▲ ▶ ▼ ▲ ▶

SPACE JUNK

by Judy Donnelly and Sydelle Kramer

Old satellites, rocket boosters, and other
debris from the space program have created
an orbiting junkyard in space.

TIME: *the present*
PLACE: *a few hundred kilometres above Earth*

You're travelling through space. All around you
is a black sky that never grows light. In the distance
you see the moon, the sun, and the stars. Beneath
you, Earth is softly glowing, spinning silently in a
blue-green haze.

But suddenly a flash of light dazzles you. It's
getting brighter and brighter—it's hurtling right
toward you. You and your fellow astronauts watch
it approach, unable to move. What could it be? An
alien spaceship? An enemy missile? A falling star?

The light streaks closer and closer. Now
you're really getting worried. In the cabin of your
spacecraft you all hold your breath. Then someone

hits a button on the instrument panel, and on the screen in front of you, you suddenly see what the light is. You and the other astronauts start laughing. It's only a screwdriver, lost in space—just another piece of space junk, like the ones you've seen before.

Funny? Yes. Science-fiction fantasy? Not really. All kinds of human-made clutter are circling the planet Earth right now—we even know a screwdriver was once in orbit. And although astronauts today can't actually identify each hunk of trash the way they do in this imaginary scene, space junk is a real danger to them and to the world's space programs.

How did it all get up there? Why didn't anybody stop it?

When burned-out rocket stages kept circling Earth, or satellites stopped working but didn't fall from the sky, nobody worried too much about it. After all, people told themselves, space stretched on and on, like an empty ocean that never reaches a shore. When flecks of paint peeled off spacecraft or when astronauts on spacewalks dropped some of their tools, nobody thought it would be a problem.

But even though space seems to go on forever, we depend on the part of it that is closest to the surface of the earth. In the same way our air and water have grown dirty over the years, this fraction of space closest to our planet is slowly filling up with junk. Today there are millions of pieces of debris above us, from bits tinier than a

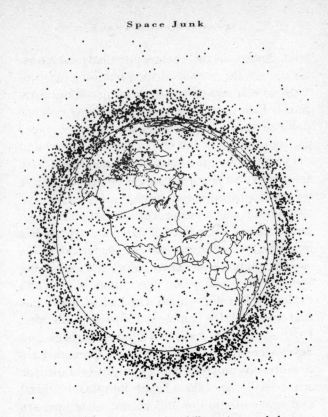

Every dot in this computer-produced illustration stands for an object orbiting Earth. About 95 percent is space junk.

thumbtack to rockets as tall as a three-storey house. And every piece, even a small one, can damage a spacecraft and put the lives of astronauts in danger.

More and more garbage is created every day. Sometimes old and useless spacecraft tumble back to Earth.

If one landed on a city, lives could be threatened. Sometimes rocket stages that have been travelling near the surface of the earth for years will explode. Chunks of metal then spread out hundreds of kilometres across our sky. Sometimes one piece of space junk crashes into another. More garbage is created and ends up travelling around the earth. Sometimes bits of debris collide with spacecraft carrying people. On one space-shuttle mission a window was damaged.

So far we've been lucky. No terrible space-junk accident has occurred. But some experts feel our luck may soon run out. An American space station might be launched before the end of the 20th century—will it ram into a hunk of junk and explode into smithereens? The United States has also sent the huge Hubble space telescope beyond Earth—will it smack into a blob of garbage and never work right again? Special weapons that are powered by nuclear energy might be built for space—will one of them smash into floating metal and rain radiation on our planet? Satellites are orbiting above us—will one crash in flames to the ground, setting fires and injuring people?

How will we get rid of all this garbage? Can scientists change the world's space program so less litter gets left behind? What are experts doing to protect spacecraft from space junk? Can we even locate all the dangerous debris up in the sky? So much of it is swimming around us that astronomers fear they might mistake bits of it for distant new stars!

In the next few years there will be even more trash. It can threaten the lives of those who soar into space. One day the biggest pieces may threaten those who stay behind on the ground.

If we want to stay free to explore the universe, space junk is a problem we are going to have to solve. After all, there's no space sanitation team to come to the rescue. Does this mean the garbage above us will haunt Earth forever? What are we doing to stop the pollution? Can space ever be the same now that earthlings have reached it? The future of space exploration may depend on how we answer these questions.

▲▼▼▲▼▶▼▲▼▶▼▲▼▶▼▲▼▶▼▲▶

CAREERS IN SPACE

from *TG Magazine*

Interested in a career that's out of this world? Look no farther than space!

Just like the limits of the universe, the possibilities are endless for work in space-related fields. And if you're a *Star Trek* fan, a night-sky watcher, a video-game player, or you simply wonder what's out there, there are lots of opportunities.

It's even possible these days to set your sights on becoming an astronaut, although it isn't easy. It means three to five years of pre-flight training. Even to be chosen for the program usually demands a university degree (or two or three) under your belt in the area of science, medicine, or engineering, plus the ability to speak more than one language, outstanding professional experience, physical fitness, good communication skills, and a lot of luck!

But you don't have to leave good old planet Earth to work in space. There are many opportunities to enjoy the endless reaches of the universe while still keeping your feet on the ground.

Consider space law. Or how about putting a psychology degree to use studying human performance in space? A flair for acting, public relations, photography, and an interest in space may lead you straight to a planetarium for work in promotions, visitor services, or star-show productions.

Naturally, studying science is the more direct, tried-and-true route. Careers in physics, engineering, biology, microbiology, chemistry, medicine, computer science, and astronomy can all lead you into space! Doctors are being sent into space aboard the shuttle to learn about the effects of zero gravity. Physicists are wrestling with unlocking the secrets of the origins of the universe. Engineers are hard at work designing a roving space robot which will be the arms and hands of the Space Station, the largest international development project ever undertaken. Computer programmers are interpreting messages sent to Earth by satellites. Biologists are studying the behaviour of animals and plants in space. And astronomers are studying the stars and planets.

How to choose from all these options? Look now. At high school, college, or university, consider space-related extra-curricular activities. Go to space camp. Visit a planetarium. Look at those magnificent northern lights, especially if you live in Whale Cove or Akulivik. Join an astronomy or a science club.

Just recently schools are beginning to offer tailor-made programs for students pursuing careers in space. Researchers, engineers, and technicians in industry, universities, and government are needed to maintain the technological base of the space program, and to further develop new capabilities in artificial intelligence, spacecraft testing, space environmental effects, and other areas.

So, what are you waiting for? The sky's the limit!

When Art Meets Space

"I don't do windows, and I don't do black holes."

So says Sharon Mavko, an artist at the Edmonton Space and Science Centre.

But she's joking. In fact, she has done black holes.

"I actually learned to create some very good ones."

Mavko also finds herself drawing comets, trying to create a scene of the surface on Mars, and even airbrushing a night scene of a car's headlights with a UFO looming overhead.

Why? She is one of an artistic in-house team at the Planetarium that prepares between two and three shows a year. Three slide projectors and some special-effect projectors beam these light-and-sound shows onto the 360-degree inner dome of the Star Theatre. Visitors lean back in their chairs and are transported into the dark regions of space.

"Creating the panoramas so that people feel that they're sitting right in a landscape is one of the most challenging and exciting parts of my job."

Each show takes between three and four months to prepare and it's all done by Planetarium staff. Writers create a script, Mavko does all the artwork, photographers shoot her work onto slides, an audio-technician produces the music, and a producer puts it all together.

The quality is high for the shows produced at the Edmonton Space and Science Centre. A number of them have been sold and played in Europe and the United States.

"It's satisfying to know my work is being seen in other parts of the world," Mavko says.

Her job is one that demands creativity, stamina, and a wide range of artistic talents.

"We have a couple of astronomers on our staff whom I consult for accuracy when I'm researching the star shows, and when I'm not researching I may be painting, airbrushing, or doing computer graphics. For one of the children's shows, I designed some hand puppets and we filmed a little animation sequence using them."

Mavko majored in art in high school, and then majored in advertising in college.

She laughs, "At one time I didn't even know where the Big Dipper was." But every day she learns more about space—and enjoys it.

Turning Numbers into Star Maps

Some people look out at space to see what's there.

Not Diane Parchomchuk. She's one of a 35-member team at the Dominion Radio Astrophysical Observatory near Penticton, B.C., who *listens* to space.

When Parchomchuk arrives at work every morning, the listening is actually over for the night. Seven large dishes organized along a railroad track, with antennas pointing at the same point in the sky, have been receiving a very weak signal during a 12-hour period. For 12 nights in a row, the information is gathered, put into a computer, and stored.

"I'm using a computer program which allows me to write software for displaying data in various forms. Basically, I turn what we've been listening to over a period of time into a picture."

The maps are then distributed to professional astronomers who use them to enhance their own studies.

Years after getting a B.A. in science, Parchomchuk completed a certificate course in computer processing at the University of California. When she returned to her home in the Okanagan Valley, her eyes turned naturally skyward—to the Observatory.

Parchomchuk's on-the-job training over her four years at the Observatory has proved invaluable. Parchomchuk admits that the work is very challenging.

"I always get to work and learn something new. The wonderful thing is, there's never a dull moment!"

2

Computer World

▲▼▼▲▼▲▼▷▼▲▼▷▼▲▼▷▼▷▶▶

SHOE

by Jeff MacNelly

▲▼▷▼▲▼▲▼▼▼▲▼▼▼▲▼▷▶▶

▲▼▶▼▲▼▶▼▲▼▶▼▲▼▶▼▲▶

WHAT'S YOUR TECHNO-TYPE?

by Carolyn Leitch

Computers have moved into offices, classrooms, and homes, but not everyone is completely comfortable with the technology.
What about you?

Dell Computer Corp. believes in techno-typ-
ing—identifying customers by the way they relate
to new technologies. It hopes this will help in
designing machines for narrower groups of users.
Its categories of techno-tolerance:

- **Techno-wizard:** A technology expert or hobbyist
  whose greatest concern is losing the edge.
- **Techno-to-go:** Wants a computer ready to go
  right out of the box. Interested in what a computer can do but not how it works. Greatest concern is being left alone without the hand-holding
  of a technical support person.
- **Techno-boomer:** Wants to look smart, and
  researches heavily before making a purchase.
  Worries about making the wrong decision.

- **Techno-teamer:** Computer is plugged into a network at the office. Big worry is network failure.
- **Techno-critical:** Relies on computers for sophisticated tasks essential to getting the job done. Major fear is loss of business from system failure.
- **Techno-phobe:** Rejects technology outright.

▲▼▶▼▶▲▼▶▼▶▲▼▶▼▶▲▼▶▼▶▲▼▶▼▶▲▼▶

WEARABLE COMPUTERS

from *The Futurist* magazine

These computers are not just portable, they're high fashion!

It's the year 2005. A car runs a red light and crashes into another vehicle. One driver appears seriously injured.

An ambulance arrives. An emergency medic quickly checks the driver's vital signs and injuries with a hand-held "track ball" containing special sensors and a video camera. The vital signs are displayed in the medic's goggles and the information is transmitted to the hospital. Meanwhile, the medic dictates comments on the driver's condition into a microphone attached to the medic's goggles. The data are matched against a medical encyclopedia on CD-ROM for a possible diagnosis and treatment.

A newspaper reporter arrives and begins entering her story into the "lapbody computer" hanging from her shoulder. A shop owner nearby, who witnessed the accident, gives a statement to a police officer wearing a "portable office," which faxes the statement to police headquarters. The shop owner then resumes taking inventory, using an optical scanner worn on his forearm to read bar-code information.

The foregoing scenario features the wearable computers now under development at the NEC Corporation's Advanced PC [personal computer] Design Center in Tokyo, Japan. The goal is "personal environment design; that is, architecture for the human environment," says design supervisor Hideji Takemasa. "We try to move our designs closer to the human being and blend the machine with the body."

The designs also blend high function and high fashion. As computers become more pervasive in everyday life, they may soon be considered fashion accessories, although very useful ones. NEC suggests that the computer store of the future will be more like a designer boutique and less like a warehouse.

Tomorrow's computer user will be wearing "whatever is comfortable for the human body," according to Takemasa. "One of the most exciting aspects in the development of these computers is that there are no preconceptions as to how they'll look or work."

Thus, wearable computers can be tailor-made for specific user groups, such as the TLC (Tender Loving Care) PC for emergency medics and ambulance drivers described in the scenario. Because this computer is worn across the shoulders, it doesn't obstruct downward vision or interfere with driving. The TLC model also frees both hands for medical treatment.

NEC expects to begin marketing wearable computers before the turn of the century.

▲ ▶ ▼ ▲ ▶ ▼ ▲ ▶ ▼ ▲ ▶ ▼ ▲ ▶ ▼ ▲ ▶ ▼ ▲ ▶ ▶

AN INTERVIEW WITH COMPUTER INVENTOR STEVE WOZNIAK

by Kenneth A. Brown

In this age of the personal computer, Steve Wozniak is one of the pioneers. He created his first computer at age 13, and then in 1977, he and his high-school friend Steven Jobs founded Apple Computers. Wozniak was the thinker behind Apple, and Jobs was the marketer. By 1980 their sales topped $100 million.

Wozniak left Apple in 1985. He continues to work as an inventor, saying there is no end to the possibilities!

Was there anyone in your background who really encouraged you to become an inventor?

Tom Swift[1], no question. He made it a good thing to invent and be a scientist. Winning science fairs at an early age or even entering science fairs was also important. If you enter a science fair and do something well, you get a lot of positive feedback from parents and teachers and the like. In your head, being an inventor becomes a good thing.

My father also gave me some direction because he was an electrical engineer. When the right times came around, he helped me with electronic projects. He would stand at a blackboard for no reason at all and teach me about transistors. He helped me learn things that weren't even taught in school. He also gave me some of the first books on computer programming, even though he didn't program himself. So, he influenced me a lot and gave me direction. I kind of wanted to be an engineer like my father.

What do you consider your first invention? Was it the Apple [II]?

Not really. Whenever you solve a problem, whether it's mathematics in school or an electronics project, in a way you're inventing. Before I worked on the Apple, I worked on some electronics problems that were really kind of inventions. I built an adder-subtracter in the eighth grade, and I solved some circuit problems such as how to build a gate[2] with two diodes[3]. The two diodes wouldn't work, so I had to put transistors in there as well.

While I was in college, I invented my own version of a blue box[4]. In the three years just before the Apple, I totally designed a video terminal and a version of the video game Pong, which influenced some of the people at Atari who wanted to hire me.

I also designed a movie system for hotel-room televisions back in 1970.

The early days of Apple have been described as a "ride on a rocket." Were you ever surprised at the way the personal computer market took off when the Apple was introduced?

No, never once. I designed these computers to show off to my friends; I didn't have any plan to start a company. I knew from my electronics background that computers were going to sell at least a million units, even when they had sold only 20 000. I used to have a ham radio licence, and I knew there was a large market for ham radios. And I also knew there were more computer people than ham-radio operators.

I didn't think we would sell as many as we did or that the industry would become as big as it is today. If I had, I would have based all the decisions on building a product that consumers would like and that we could sell. And, we probably would have made the wrong decisions technically and built the wrong product.

What do you think it takes to come up with a good invention? You mentioned that everything converged for the Apple II. Is it just luck?

You've got to have a pretty darn good idea in your head of an end goal. You can't just sit down and start using some tools you were taught and see where it takes you. You need one goal, and your goal has to coincide with something that somebody else wants to buy or something that will save them money.

Are there any particular traits that an inventor needs to have?

It's very good if you can spread yourself over several disciplines.

Some people get down to one discipline. For example, I designed a certain part of a tape-recorder amplifier; I took that as my job. I like to do these one after the other, maybe 10 a year.

But in my experience, it has been very motivating to be able to whip out a piece of software for this task, build up a language over here, connect these chips together. When you can transcend different disciplines, several disciplines, you can make things much more optimal. And this is what makes something artistic. You make a better circuit when you know what kind of code is involved.

Getting good feedback is also important. I had the best in the world when I was developing the Apple II. When I had something, I would just take it down to the Homebrew Computer Club in Palo Alto [California].

I was shy. I was really shy. I would never raise my hand and say, "I have something I'm going to be showing." I would just set it up on a table, and a few people would gather around me by accident. I developed my own group that way, and I would just tell them what I had come up with. And the look in their eyes was the sort of feedback you don't quite get from a boss in a company. They knew that what I had done was important, and they knew why it was good.

Negative feedback is also a problem, isn't it? One inventor I talked to said he never tells anyone what he's working on because people can be so negative. Have you ever felt that way?

Exactly, exactly! Right and left! Right and left! And I put up with it. I'm really patient and nice to people. But it's hard when you have an idea and you want to implement it the way you see it and everybody else tells you why it won't work. The funny thing is that sometimes they'll have very good logical reasons and they are right from some viewpoint.

But the trouble is that often a different approach makes no difference in the outcome. For example, I can use approach A, approach B, or approach C—it doesn't matter which I choose. In the end, the important thing is that I get it done. And in getting it done, the most important thing is confidence.

Anybody can point out a minus— "Well, that's not the way because it has a higher density of electrons," and this and that. The trouble is that it's not a fair way to judge an idea or an invention. Other people will look at a new idea and say, "Wow, that's great. Show me more later."

Do you usually work alone?

Anything that I've ever been proud of or was acknowledged for later was always done on my own. As a matter of fact, in school I was always very much an individualist. To me, a teacher or a classroom didn't matter at all. I learned by reading in my bedroom or dorm room late at night or whenever I wanted to. Sometimes, I would do a whole course in a two-week cram.

Alan Baum is about the only other person I've done computer work with in my life. He and I went back and forth on some of the Apple II design. He would suggest a direction, suggest some code to start it with, and I would do some improvements on it because I was a good coder. We would work together like that. But that was about it.

What do you think the future holds for inventing?

Are you referring to the idea that it's all been done before? Oh no, no, no, no! It always seems like so much has been done, but I'm sure people thought that way in Rome thousands of years ago. I've never had that impression. There will always be new inventions simply because there's a need inside us to express our creativity and inventiveness.

1. **Tom Swift:** "The Tom Swift, Jr. books were a series about an inventor that was popular when I was growing up in the 1950s. They were sold alongside

Nancy Drew and The Hardy Boys. I can recall always waiting until a new one I hadn't read yet came out!"—*Steve Wozniak*

2. **gate:** a device that outputs a signal when specified input conditions are met

3. **diode:** an electronic component consisting of a semiconductor and two attached electrodes

4. **blue box:** an illegal electronic device that generates tones to "fool" and trigger telephone circuits

▲▶▼▶▲▼▶▲▼▶▲▼▶▲▼▶▲▼▶▲▼▶▲▼▶▲▼▶▶

PROGRAM LOOP

by Jill Paton Walsh

When Robert's parents leave him home alone, he gets involved with a very unusual computer.

The summer he was sixteen Robert's parents left him alone in the house for three weeks. His father was going to New York on business, and his mother needed a holiday, and wanted to go with him. Robert had had glandular fever; think of something for a boy to do, and he couldn't do that, because he needed rest. He just had to stay put and bear it. His mother stocked the freezer lavishly, and wrote instructions on postcards taped all over the house, before getting on the plane, a day before his dad, through some mix-up over tickets.

"Even convicts don't have solitary confinement except as a punishment," said Robert, but only to himself. Then, on the last day, Robert's father bought him the computer.

"I thought it might help to while away an idle hour," he said, putting the machine, leads dangling, on the kitchen table.

Robert's eyes widened. "That's brill[1], Dad!" he said. "Mega amazing! But can we afford it?" In spite of trips to New York, Robert's father worked in a field that was all glamour and not enough cash.

"There's a lot more of it," his father said. "Give me a hand getting it in from the car. Where do you want it?"

There was indeed a lot more of it. A very good monitor, better than a TV set. A disc drive. A box of discs. A printer. Fanfold paper. Leads, the user manual. The disc drive was huge; 800K.

"But can we afford this?" Robert said, as they laid it out on the desk in his father's den.

"For my boy, nothing but the best!" his father said. "No, truth is, Robert, it's all second-hand, and I picked it up very cheap. Astonishingly cheap. As if the man wanted to sell it to me in particular. So it's all yours, and don't worry about the money."

"Imagine having all this, and wanting to sell it!" said Robert. "It looks new."

"Well, I didn't actually meet the guy who had used it. Bit odd, in fact. He seems to have disappeared, and his father is selling up. Our good luck. Do you know how to use all this?"

"I've never had my hands on one of these before," said Robert. "It's very new and advanced. Supposed to be the absolute best. But I expect I can manage."

"Good. Good. Well, I've got to pack."

Robert spent half an hour getting all the cables plugged in and everything connected together. To connect things to the mains[2] he needed one of those floating sockets which takes four or five plugs, and needs only one plug in the mains; there was only one power point in his father's den—the one his mother used for the vacuum cleaner. And he didn't like to go and buy one till he had seen his father safely off—it didn't seem friendly.

So one way and another it was the following day before he really settled down to work on it, and he had already had a long evening in front of the television set, and the amazingly dislikable experiences of going to bed without anyone to say goodnight to, and going out shopping without anyone to whom to say "Just popping down to the shops...." He had every intention of getting lost to the world with his new machine.

Getting it up and running proved easy enough. It might be secondhand, but it was in perfect order. The previous owner had left not so much as a smudgy fingerprint on a key, never mind any little snags to correct. Robert summoned up the BASIC[3] he had learned at school, and began to play. First he made it print HALLO ROBERT YOU GENIUS on the screen, flashing on and off. Then he made it draw the diagram for Pythagoras' theorem. Then he started on his own great project—writing a program to make it play Bridge. This was really an interesting job. You had to make the machine divide fifty-two signs, one for each card in the pack, into four "Hands" completely at random. Robert got a bug in his program, and the machine kept making hands which repeated cards in one of the other hands. Obviously, he had to sort that out before he taught it scoring and bidding....

Eventually he realized he was not just hungry, but *ravenous*. It was six o'clock, almost suppertime, and he hadn't bothered with lunch. No wonder. No mum, bringing sandwiches, or saying, "Robert, you must eat." Robert winced. He honestly hadn't expected to miss his mother. He had expected to be glad to see her when she got back, but actually missing her.... Oh, well, live and learn.

But he was in a dilemma. The machine was quite warm now, having been run all day, and Robert thought it would be better to turn it off while he made and ate supper. But if he turned it off he would lose the Bridge program unless he managed to "Save to disc" and he hadn't yet mastered the disc drive. But he was very hungry. He switched the drive on, and put one of the old owner's disc's into it. "Save 'Bridge' to disc" he typed.

DISC FULL came up on the screen.

Robert scrabbled through the manual, and typed "*Enable. *Erase."

ERASURE PROHIBITED said the screen.

"Hell!" said Robert, and tried another disc. The third he tried seemed to be empty, and he stored his program safely on it, switched off, and got himself into the kitchen. A pork chop? No, takes time to cook, and doesn't look too good to eat raw. He concocted a sandwich of pilchards[4] and peanut butter, topped it up by eating a whole drum of ultra-rich ice cream, and promptly fell asleep on the sofa in front of the television, until the piercing close-down tone woke him up. He seemed to have indigestion, so he made himself hot cocoa, using milk which had been standing on the doorstep all day and tasted a bit funny. The cocoa had floaters in it, but it was warm, anyway.

At least he was learning the value of breakfast. The Bridge program kept him going through lunchtime day after day. He needed books, too. He went to the computer shop to buy them. And while he was there he noticed on a price list for discs, the brand name, unknown to him before, of the ten discs that had come with the machine. Eight pounds fifty each! More than twice as expensive as any others.

"Lucky you," said the shop manager. "Those are the very best. They can be reused forever. Just erase them, and you've got all you need for all the programs you can write."

Just erase them. But he kept getting the ERASURE PROHIBITED message. He went through the routines in the manual, meticulously, checking every step, and still the message came up.

Oddly, it was only after hours and hours of struggle to erase that he thought of loading the used discs, and seeing what was on them, so carefully protected. FILES LOCKED. ACCESS

PROHIBITED said the screen.

Robert swore. Then he went downstairs to the kitchen and made himself a large stack of sandwiches, and put them on a tray together with the electric kettle and the jar of instant coffee, and a bottle of lumpy milk, and settled down grimly with these supplies to do battle with the devilment of the previous owner of the machine. After some time he got a new message on the screen. It now read ACCESS CODE? The little green dot bleeped at him, waiting. If you knew the code, you typed it in, and then the thing let you read the disc. The chance of guessing the code was zero.

A disassembler. Perhaps with a disassembler he could break the code, or read it rather, finding it in the intricacies of the program....

He bought a disassembler with the last of the holiday money his parents had left him, and ate two large meals from the freezer, to stock up on. Somehow he had gone off sandwiches. There weren't any clean socks left in his drawer. He put on a pair of sandals, and when that left his toes freezing, he brought the electric fire from the spare bedroom and beamed it at his feet. He was good at computers. Very good. But this was a pig of a task. More days. In desperation he read one of his mother's postcards, ran the washing machine, and put on clean socks, damp.

And then at last he got there. The code was RRT841. Odd, that. His own initials, Robert Randall Thompson, and the date—'84, and 1. Rubbish.

Coincidence. "All this living alone is driving you loopy, Robert lad," he told himself, and typed. ACCESS CODE? said the screen.

Triumphant, Robert typed in "RRT841."

On the front of the disc drive, the little red light lit up, and the machine purred quietly. He'd got it!

Got what, exactly?

He'd got a message that said, DISCONNECT VDU[5]. CONNECT TELEVISION. Nothing he did would remove this silly message, or persuade the machine to disgorge more of whatever was on the disc.

In the end, in desperation, he did it, using the small portable TV from the ironing room. His mother hated ironing, and television, but somehow found them more bearable both together.

He set the TV down, and plugged it in. At once the screen faded out the CONNECT TV message, and said, AT LAST. I THOUGHT YOU'D NEVER MAKE IT. WELL, OBVIOUSLY, I KNEW YOU WOULD EVENTUALLY, BUT IT DID TAKE YOU LONG ENOUGH!

Robert looked at this message for a long time. It made a nasty prickling sensation run down his spine. It certainly didn't look like any other computer message he had ever seen.

After a while it blipped off. He felt a momentary deep relief, and began to convince himself that damp socks and bad eating caused delusions in post–glandular-fever sufferers, when the screen came up with another line. FOLLOW DIRECTIONS ON DISC.

The chill returned. Robert thought deeply, and typed in, "I am not who you think I am. Previous owner of machine gone away."

The screen said YOU ARE WHO I THINK YOU ARE. FOLLOW DIRECTIONS ON DISC.

Robert typed in "No."

The screen showed BUT I KNOW THAT YOU WILL. WHEN YOU ARE READY, TYPE "CHAIN SHOPPING LIST." THE DISC WILL SHOW YOU WHAT YOU ARE TO BUY.

Robert switched everything off, and went for a walk. But the message reappeared when he switched it on again. He could rid himself of it easily enough, by using the VDU instead of the TV so he could settle down to his Bridge program, but of course the mysterious message nagged away at the back of his mind and eventually he gave in, and typed "Chain Shopping List."

Purring, the disc drive showed him a list of computers. Model numbers, brand names. More or less everything on the market in the way of micro-computers. Some of the items had BOUGHT AND SHIPPED against them. At the foot of the list, when the machine scrolled down to it, it said, BUY ALL ITEMS NOT MARKED BOUGHT AND SHIPPED.

Robert typed in, in a fury, "Don't be bloody ridiculous, I can't buy all that stuff!"

WHY NOT? said the screen instantly.

"No money" typed Robert.

OPEN A BANK ACCOUNT said the screen.

Rage shook him. He was hating this: hating being bossed about in this ridiculous way, hating himself for not being able just to switch off and ignore it. But one reason why his mother hated television was that neither Robert nor his father could ever bring themselves to switch off lousy programs....

Seething, he typed in, "I'm too young. I can't do that."

The screen answered so fast it took his breath away: YOU WILL FIND THAT YOU CAN.

And of course, he did find that he could. He told the bank manager an elaborate fairy story about being alone in the house, about some bill needing paying urgently, and his father being about to put a cheque in the post, and how he, Robert, would need a paying-in book to pay it in, and a cheque book of his own to pay the bill....

The bank manager was amused. "There are easier ways of doing this, you know," he said. "I'm surprised at your father. But perhaps he'd like you to get used to banking early. Most people leave it till they have to handle student grants. But we like to get customers young. Just let me know if I can help in any way; if the letter with the cheque doesn't arrive promptly, or anything. How much are you putting in now?"

"Now?" said Robert, alarmed. "I haven't got any till my dad sends this cheque...."

"Well, it isn't usual to start a bank account

with absolutely nothing," the manager said.

"I've got fifty p[6]," said Robert.

"Great oaks from little acorns grow," said the manager cheerfully.

He was quite right, too. When, three days later, Robert used his new card to request his bank balance from the automatic till, under instruction, of course, from the inexorable screen, the balance shown was £100 000.50.

It was fun, in a way, marching into computer shops buying wildly expensive things. They kept ringing the bank to check the money, and seemed very respectful when they came back. Just the same, he began to worry that all this activity might draw attention to his account, so he withdrew a huge amount in cash, remarking to the desk clerk that his father needed it to pay builders, and began to pay cash for things. He listed into the computer what he bought. The boxes piled up in the sitting room. And the screen talked to him. It said ZAPPO! when he got some prized item on the list. Sometimes he couldn't get what it wanted, the listed models were superseded, and he had to ask for further instructions. Once he couldn't get the right thing, because the code number was too high; the computer listed a Diogenes 800, and the shop said it didn't exist. There were only two Diogenes micros: 50, and 100.

EVEN BETTER said the screen. BUY THE EARLIER MODEL. Odd, that.

Robert began to talk back a bit to the screen.

"Whoever wants all this stuff?" he asked it. "Nobody uses all this. It isn't compatible. You choose one system and get only what goes with it."

I AM A COLLECTOR said the screen.

"Nobody collects computers" said Robert.

I DO said the screen. I HAVE LOVED THEM EVER SINCE MY FATHER BOUGHT ME ONE OF THOSE FIRST MODEL B'S WHEN I WAS A KID.

The day came when Robert had to put a box in the hall, there being no room left in the sitting room. He wondered what his mother would say if she saw the mess, and then he looked at the calendar, and realized it was only three days till his parents came back. He bolted up the stairs, and asked the computer how to start shipping the stuff.

THAT'S THE BEAUTY OF IT said the screen. IT ISN'T A PROBLEM. YOU JUST KEEP THE STUFF.

Keep it? Oh, gods....

"Impossible. Awaiting further instructions" he typed.

AS ABOVE said the screen.

"Look, what happened to the stuff the other guy bought before he quit? Can't we do the same with all this?"

THE PREVIOUS PURCHASE AGENT GOT TOO CLEVER said the screen. HE GOT HIMSELF SHIPPED WITH HIS LAST CONSIGNMENT. YOU CAN'T IMAGINE THE TROUBLE IT CAUSED. DON'T EVEN THINK ABOUT IT.

"I think it's time you came clean with me"

typed Robert. "Exactly where is all this stuff going?"

NO MARKS FOR GUESSES. NOT WHERE, WHEN. THE PREVIOUS AGENT GOT OVEREXCITED ABOUT IT. I HAD IT ALL SET UP NICELY: A SYSTEM FOR SHIPPING ANTIQUES FROM YOUR TIME HORIZON TO OURS AND THE IDIOT CRATED HIMSELF UP WITH A LOAD OF IBMS AND CAME TOO. THE BUREAUCRATS WERE OUT-RAGED. HE JUST DISAPPEARED THEN AND ARRIVED NOW WITH NO PAPERS, NO PAST, NO CLUE HOW TO BEHAVE, NO MONEY ... THEY HAVE ABSOLUTELY FORBIDDEN FURTHER SHIP-MENTS. I WAS DISMAYED. THEN I THOUGHT OF YOU. PERFECT. NO SHIPPING NEEDED. YOU JUST KEEP THE STUFF.

"But I can't!" typed Robert. "There isn't any-where!"

ALL WORKED OUT. PUT IT STACKED CLOSELY ON THE BOARDED PART OF THE ROOF SPACE, BEHIND THE COLD WATER TANK. COVER WITH BLACK POLYTHENE SHEETING AND STRING. TIED DOWN AS IF ANOTHER TANK. THEN FORGET ABOUT IT.

"And someday I meet you? Is that it? You're crazy. What if we move? What if I die young? What if I won't part with the stuff when you suddenly turn up and ask for it?"

YOUR QUESTION A, said the screen, YOU DO NOT MOVE HOUSE. YOUR QUESTION B YOU DO NOT DIE YOUNG, THAT I DO KNOW. YOUR

QUESTION C NO PROBLEM. YOU STILL HAVE NOT APPRECIATED THE BEAUTY OF THIS ARRANGEMENT.

"I certainly haven't!" typed Robert. "What the hell do you mean about antiques?"

ANTIQUE EARLY COMPUTERS IN MINT CONDITION said the screen. PURE AND PERFECT DELIGHT. THERE WERE SO MANY KINDS IN THE EARLY DAYS. YOU DO NOT FOLLOW? THINK ABOUT LOOPS.

Robert did think about loops. He thought about them while he heaved and stacked boxes; while he struggled with the loft ladder, and polythene sheeting, and string; while he lay in the bath soaking off the dirt of the roof space, and the deathly weariness that the heaving brought on; well, he was supposed to be resting after glandular fever… He thought about them while he chased more knowledge through the user manual, and the disassembler program….

His kit was networked in some crazy way. Networked into the future computer that gave orders. With some ingenuity he extracted from his disc a code for user directory, and from the directory the locking code for the user issuing the purchase list. It was RRT 20491. Robert Randall Thompson, in two thousand and forty nine, no doubt! Of course the shipping would be no problem; the bossy purchaser was himself, grown old in fiendish ingenuity. If he kept the things, he would own the things … a unique collection of antique computers in mint condition!

But what about that figure one? What was that doing? Robert learned about loops. The computer could be sent round and round loops in the program; it could count the number of times it went round them. 1 meant it was going round for the first time, but the very presence of a number meant it was going to go round more than once....

It mustn't. He couldn't bear it. He set his tired wits at the problem again. This particular thing wasn't difficult. The computer was set for six loops. They were nested; if he touched a computer again it would find him when he was twenty-four, thirty-two, forty... Each time the loop was shorter, one inside another....

He found the program line that counted the loops. Easy. This would be easy. He inserted a line. "If n is greater than 1, endprocedure" he wrote. And then, "run."

The screen cleared. He had wiped out the loop, and the program with it. The house seemed suddenly empty again, and full of relief.

He waited for his parents to come home before he investigated. Somehow their being around made him feel safer. He found he had a bank account with fifty p in it. Oh, well, that would be handy when it came to student grants. There was still a pile of kit in the attic, behind the tank. He wondered if he should drag it all downstairs again sometime, and dump it in the river....

Then he thought, he's incredibly ancient by

2049. He's probably long past girls and drink, and any kind of fun. If the micros give the poor old geezer pleasure, where's the harm?

So he left the stuff exactly where it was.

1. **brill:** short for "brilliant" (slang)

2. **mains:** circuits that carry the combined flow of secondary branches of a system

3. **BASIC:** Beginner's All-purpose Symbolic Instruction Code: a simplified language for programming and interacting with a computer

4. **pilchards:** small, herring-like fish

5. **VDU:** visual display unit (computer screen)

6. **fifty p:** fifty pence; worth a little over Canadian $1

3

High-tech Entertainment

▲▽▷▽▷▲▷▽▷▲▷▽▷▲▷▽▷▲▷▷▷

THE FAR SIDE

by Gary Larson

Hopeful parents

▲▽▷▲▷▽▷▽▷▲▷▽▷▲▷▷▷

▲ ▼ ▶ ▼ ▶ ▲ ▼ ▶ ▼ ▶ ▲ ▼ ▶ ▼ ▶ ▲ ▼ ▶ ▼ ▶ ▲ ▼ ▶

THE BUNGEE LUNGE

by Karen McNulty

Here's your giant rubber band. Now jump! It's only a 10-storey plunge—and science will spring you back.

The Science Behind the Bounce

Ready?
When standing high on a jump platform, you have lots of potential (stored) energy.

Jump!
Leap off and your potential energy is converted to kinetic energy, the energy of motion. For a few seconds, you experience free fall, until there's no more slack in the cord.

Stret-t-ch
Then the cord starts to stretch. This stores the energy of your fall in the cord.

Bounce
This stored energy springs you back up. You fall and bounce again ... and again ...

Phew!
Each bounce disperses some of your energy,
so eventually you stop. You'll have to hang
around until someone lowers you to a raft or
the ground.

You're hanging onto the railing of a bridge, 46 m
above a river. Your friends on the bank below seem
awfully small; looking at them makes you dizzy.
Someone standing behind you is counting down
"three … two … one!" Defying every sane notion in
your brain, you leap—headfirst.

The 100-km/h fall toward the water terrifies
you. But just as you close your eyes for the icy
plunge, something happens: You bounce back!

Better thank your lucky *bungee cord*—that
wrist-thick band of latex rubber strapped to your
ankles and anchored to the bridge. Because it was
the right length, it kept you high and dry. And
because it stret-t-t-ched and recoiled—giving you a
few good bounces—it used up the energy of your
fall so you didn't get torn limb from limb. Phew!

Those who have done it say it's the thrill of a
lifetime—"a natural high." Others call it crazy. But
everyone knows it as "bungee jumping," the sport
springing up (and down) across the nation.

At least one group of people has been
"bungee jumping" for ages: the men of Pentecost
Island in the South Pacific. They make cords from
elastic vines, lash them to their ankles, and plunge
off wooden towers into pits of softened earth. For
these islanders, jumping is a springtime ritual,

meant to demonstrate courage and supposedly ensure a plentiful yam harvest.

In North America, jumpers take the bungee plunge just for the excitement of it. Scott Bergman, who runs a bungee-jumping company in California, explains the appeal. "It's a feeling of having absolutely no control—and loving it."

And it doesn't take any skill. Just $75 to $100 and *faith*—in physics. It's a simple physics equation, after all, that lets "jump masters" like Bergman determine how far the cord will stretch when you take the plunge—and whether it will stretch too far.

Weighing the Odds

The major variables are the stretchiness, or *elasticity*, of the cord—predetermined by the manufacturer—and the jumper's weight. As you might guess, "the heavier you are, the more the cord is going to stretch," says physicist Peter Brancazio.

By weighing customers (they don't just ask), using the equation, and adjusting cords, jump masters have bounced thousands to safety. (There *have* been some deaths—usually caused by frayed cords or other faulty equipment.)

Jump experts can even adjust the cords to give their clients custom-made thrills. "When we jump off bridges in California," says Bergman, "we ask the people if they want to just touch the water, dunk their heads in, or go all the way. We can really get it that exact."

Really? "I wouldn't trust them," says Brancazio, "but I guess they can."

If, for example, Bergman calculates that you'll crack your skull on a rock in the river, he can shorten your cord. "That starts the stretch at a higher point off the ground," says Brancazio.

Or you can jump with two cords. "In that case," says Brancazio, your weight is "equally divided between the cords so each stretches half as far."

Chances are, you'll scream just as hard with fear and delight.

▲▼►▼▲▼►▼▲▼►▼▲▼►▼▲▼►

THE THRILL OF THE THREAT

If you ever go bungee jumping, chances are you'll experience a "rush": Your heart will race, your breathing will deepen, and you'll feel a super energy charge. It's a more intense version of what you feel when a big dog barks at you or you have to call someone for a date.

If you add the dryness in your mouth and the butterflies in your stomach, you'll remember that it isn't a wholly pleasant set of sensations. But the reactions that cause them do serve you well in "threatening" situations. They give you the extra oxygen and energy you might need to run away or fight for your life. Scientists believe this "fight or flight" response evolved to help us stay alive.

It all starts when your brain senses stress or excitement and instructs your body to pump out *hormones*. (You probably know the name of one: *adrenaline*.) These messengers carry the chemical signals that tell your heart and lungs to work faster and your cells to release more energy. You really do get a charge, whether you need to use it or not.

You also get a strange sense of elation. That's because stress stimulates your body to make its own "feel-good" chemicals, called *endorphins*. Endorphins are potent painkillers, which may explain why you don't feel the bite of the bungee cord, why some athletes can continue to perform despite injuries, and why anyone would even consider running a marathon.

TALKING TO THE AIRWAVES: INTERACTIVE TV

by Lila Gano

Sit up, couch potatoes! You can't just watch the set any more—interactive television demands viewer participation.

In the past, television was a one-way communication device. Electrical signals travelled to homes with no response from the audience. Interactive television promises to make TV viewing a two-way process. Instead of sitting in front of a TV set, a viewer will be able to react to what is happening on the screen by simply pressing a button. TV sets would come equipped with a panel of buttons similar to a remote-control device that transmits audience reactions back to the television station.

The basic idea of interactive TV has been around since the 1950s. The public was first introduced to the concept by Winky Dink, a cartoon character. When he appeared on TV, children were invited to interact by outlining letters of the alphabet on a plastic overlay they placed on the screen, thus revealing secret messages. In the 1990s, newscasters sometimes ask audiences to participate in another way. When a controversial story airs, viewers are invited to call a certain number if they agree with a certain point of view and another number if they disagree. The calls are tallied by a computer at the TV stations, and the results are broadcast later.

Some sport enthusiasts are already enjoying interactive TV on a limited basis. In New York City, viewers can subscribe to a service that allows them to have control over what is seen on the screen. With the punch of a button, they can change camera angles during a ball game or select different levels of aerobic exercises. For a small monthly fee, viewers are provided with a remote-control device that signals an electronic unit on top of the TV set. This unit sends messages to a computer at the local TV station. The messages contain directions for the changes that viewers want to appear on their screens.

In California, a new interactive TV system makes baseball come alive. As the batter steps to the mound, viewers predict the outcome of the play. Their predictions enter a special computer terminal and are relayed to a computer at the TV station that keeps score. To make watching the game on TV more exciting, stations offer prizes to those who consistently make the correct choices.

In the future, interactive TV may play a big role in assessing public opinion. Viewers may be able to respond immediately to TV shows, registering complaints or praise. Advertisers could evaluate the impact of their commercials by obtaining immediate viewer reaction. When a politician makes a speech, public approval or disapproval could be quickly measured. Interactive television may someday allow citizens to vote for local and national candidates from the privacy of their homes. For physically disabled voters, this would be a great service.

Interactive TV may also have a place in the classrooms of tomorrow. In addition to the traditional approach to education, lectures or lab work could be televised. Special effects and graphics could be used to liven up lessons or illustrate complex ideas and concepts. More research and study is needed, however, before TV obtains its teaching credentials.

▲▶▸▼▲▼▶▸▼▲▼▶▸▼▲▼▶▸▼▶▸

HIGH-TECH FASHIONS

by Bernadette Morra

Just imagine if merely zipping on a ski suit could turn you into a better skier!

Today's sleek outerwear fabrics are engineered with aerodynamics in mind—a boost to racers looking to shave a few seconds off their time. High-performance materials also provide greater warmth and comfort, enabling sports enthusiasts of all calibres to stay out in the cold longer, which allows more time to improve technique.

"Advances in fabric technology have enabled a lot of people doing different activities to reach a level of success they have never achieved before," says Deborah Durham, technical expert for Polartec, a range of high-performance synthetic fleeces engineered to keep the wind and cold out, and warmth in.

Polartec has been worn by the crew of the U.S. space shuttle *Discovery*, mountain climbers, and dogsledders crossing Antarctica. Massachusetts-based Malden Mills, the maker of Polartec, was the official underwear supplier for the 1992 U.S. Winter Olympics team.

Tests by athletes wearing various Polartec products showed that "climbers were able to climb more rapidly because they were lighter and they could stay out longer because they were warmer," says Durham. But you don't have to be an Olympian to appreciate the great leaps made over the last decade in textile technology. Anyone who has been hiking, snowshoeing, even shovelling snow knows what happens when you perform any kind of activity in the cold.

In bulky wool layers we soon start to sweat, and the sweat starts to freeze, making things very uncomfortable. Today's high-tech fabrics provide a sort of climate control. When used in conjunction with moisture-wicking underwear and a water-repellent shell, the fleece system of dressing can provide complete comfort even in the worst conditions.

"The ideal would be to wear polypropylene underwear, which wicks moisture away from the skin, then a layer of fleece and a Goretex or Protex shell," recommends Dyann Mills, sales manager for Banff Designs, a Canadian manufacturer which makes jackets and pants from Polartec fleece.

"The Goretex and Protex are windproof, waterproof, and breathable. The polypropylene and fleece pull the moisture away and transfer it to the outer shell where it evaporates. At the same time, the fleece acts as insulation to keep warmth next to the body."

The fleece is thickly napped fabric formed when tiny needles pull polyester fibres through a length of jersey. "Air is trapped between the fibres during the napping process, allowing the exchange of air through the fabric so it can breathe," Durham explains, pointing out that polyester has come a long way since the stiff and clammy leisure suits of the '60s. "We're reaching the state of the art now."

Today's polyester fleece is machine washable, quick drying, water resistant, and comes in a range of weights, some of which have an antimicrobial finish to cut down on odour from perspiration.

Fleece has been used as the basis for a range of new outdoor products. Polartec's Windproof Series 1000 jackets and pullovers look like any other fleece tops, but have a layer of fine membrane sandwiched between two layers of fleece. "You can light a match and blow through the fabric and it won't blow out," Durham demonstrates. "It's completely windproof."

The jackets do such a good job of keeping cold out, there are pit zips under the arms to allow air to circulate should things get too steamy. Combining warmth and windproofing in one lightweight garment is especially appealing to cross-country skiers and cold-weather hikers.

For skiers, divers, bike racers, and others concerned with aerodynamics, Malden Mills has developed a thermal stretch unitard with a nylon/Lycra outer shell and fleece lining. The suit provides all the comfort of a regular racing suit with the added plus of a windproof, abrasion-resistant shell and a fleece lining which wicks perspiration away from the body so skin remains dry and warm.

Durham predicts that the future of fashion lies in high-tech materials. "It won't be long before we see more streetwear made out of these fabrics," she believes.

GET REAL! THE WORLD OF VIRTUAL REALITY

by Nancy Day

It's like being inside a video game. You are in an artificial world where you can interact with the computer-generated objects around you.

In this artificial world, you might see a ticking clock and decide to pick it up. You watch as your hand reaches down and grabs the clock. You bring it closer and the ticking gets louder. It is not reality. But it isn't just computer graphics either. It's virtual reality.

Virtual reality is a very advanced form of computer simulation. Already it is being used for everything from letting architects walk through houses before they are built to helping scientists study the surface of Mars. And one day it may be as much a part of your schoolwork as a notebook and pen.

Computer simulations have been used for years to create imitations of situations or objects. Simulations let people see what a real situation or object might be like. For example, pilots practise emergency landings using computer simulations rather than take risks using real planes. Engineers create three-dimensional (3-D) computer drawings of their designs to see how they will look before spending the time and money to build the objects.

Virtual reality takes computer simulation a step further. It provides a way for you to actually step inside a world created by computer graphics. Instead of just looking at the artificial world, you can explore it. You can pick up and put down objects. If you turn your head, what you see changes, just as it would if you were really there. Through 3-D sound, you can hear footsteps behind you or birds overhead. Virtual reality makes a direct link between the human being and the computer. This link lets people actually participate in computer-created environments.

Although the tools used to create this link vary, most virtual-reality systems use a head-mounted display of some kind and a special glove. Images are projected into each of the participant's eyes through a helmet, goggles, or a viewing device. The two images, created by a computer, provide slightly different views, like our own eyes. This helps to create a realistic 3-D effect. The display fills the whole field of vision. Tracking devices in the glove sense movements and translate them into signals the computer understands so that the world responds to the user's wishes. The results are remarkable. You see your hand wave the moment you wave it. As you pick up an object that you know is only a computer graphic, it seems as if your hand is really picking it up. Add a treadmill and you actually can walk through the virtual environment.

The uses for virtual reality are mind boggling. Researchers could develop new chemicals by "feeling" how the molecules fit together. People could watch 3-D movies that play along with their actions or ride a hair-raising roller coaster without ever leaving their chairs. A surgeon on Earth could perform an operation in a virtual-reality environment, controlling the movements of a robot operating on an astronaut hundreds of thousands of kilometres away. Virtual-reality-controlled robots could take scientists to the bottom of the ocean, to the surface of an unexplored planet, or even inside the human body. That's virtually amazing!

▲ ▼ ▶ ▼ ▶ ▲ ▼ ▶ ▼ ▶ ▲ ▼ ▶ ▼ ▶ ▲ ▼ ▶ ▼ ▶ ▶

VIRTUAL SKI TRAINING
by Josh Lerman

It had to happen. A "virtual reality" ski-training system has been developed that "mimics the view, movements, and suspense associated with downhill skiing," according to NEC, the system's developer.

Users wear what is called a "headmount," a pair of goggles that provides the visual sensations. Users stand on metal plates that simulate underfoot terrain sensations. Pulse monitors attached to a user's fingers measure stress, and the machine adjusts the terrain difficulty to suit the user's stress level.

VIRTUAL CATHEDRAL

from Discover *magazine*

A lost treasure of the Middle Ages has been resurrected with the help of computer-aided design software. When Cluny monastery in France was built 900 years ago, it was the largest and most beautiful church in the world, and it remained unsurpassed for 400 years (in size, anyway), until St. Peter's Basilica was built in Rome. Religious wars and the French Revolution left the church in ruins, though, and today only an octagonal bell tower and a part of the main transept[1] remain.

To celebrate the opening of a museum at Cluny, students from the Academy of Art and Technology in Paris, with help from IBM engineers, spent nearly 4000 hours of computer time recreating the cathedral. They based their reconstruction on extensive excavations and drawings made 50 years ago by the American art historian and archaeologist Kenneth Conant. Visitors to the museum can watch a video that shows how the church looked at the height of its glory.

---

1. **transept:** in a cross-shaped church, the transept is either of the two short arms at right angles to the long main part of the church

▲ ▶ ▼ ▶ ▲ ▼ ▶ ▼ ▶ ▲ ▼ ▶ ▼ ▶ ▲ ▼ ▶ ▼ ▶ ▶

DESIGNING A CITY FOR MARS

from *The Futurist* magazine

A Martian-city theme park proposed for Japan will take visitors 10 000 years into the future.

If a team of visionary 20th-century architects has its way, Earth dwellers won't have to wait 10 000 years to visit a city on Mars.

The Hyperion Project, a proposal to build a theme park in Takasaki, Japan, based on the concept of a Martian city of the future, is the creation of Michels Bollinger Architecture (the team of world architect Doug Michels, concept visionary Peter Bollinger, and architect James Allegro). Their proposal won the Prize of Excellence for pavilion design at a Young Astronauts Club (YAC) competition in Japan in the early '90s.

As the designers envision it, a city on Mars will allow visitors 10 000 years from now to stroll through gardens of mutant vegetation, celebrate the births of new species created through genetic engineering, watch robots engage in the ancient

Peter Bollinger/ Michels Bollinger Architecture

The Hyperion Project
A Martian theme park on Earth

Japanese sport of sumo wrestling, chat with killer whales, and intensify their awareness of sound by stepping into a chamber that surrounds them in absolute silence.

The Hyperion Project calls for a Great Pavilion—a single large pavilion for the theme park rather than a collection of isolated structures—which would house two integrated components of the park: the two-level Astropark, featuring rides, participatory exhibitions, and other entertainment and educational features, and Mars City, comprising hotels, restaurants, conference facilities, and other services to meet the needs of visitors.

The designers believe that the Great Pavilion concept—one pavilion rather than many—is crucial to the park's success in meeting its conceptual goal of advancing the development of human beings. The Hyperion Project aspires to help visitors integrate mind, body, and spirit, encouraging them to participate in creating the ideal image of the future human, or "Miraijinrui." In the "I Being" area, for instance, visitors will encounter their own image, symbolizing that the ideal future human species begins with each individual. The I Being Contemplation Plaza encourages introspection and coming to terms with one's place in the universe.

Among the many other proposed features and attractions of the Hyperion Project:

- **The Living Globe**, a real-time image of Earth as seen from space, projected onto the surface of a spherical screen to show moving images of envi-

ronments, weather patterns, or natural phenom-
ena.

- **Miracle 10 000**, a ride through the centuries, from past to future.
- **Dreamsphere**, a living encyclopedia of human dreams, where individuals may add their own dreams and compare the symbolism with dreams from other cultures.
- **Dance Mo*tion**, including areas for performance by Hyperion dancers and for participation by visitors in a "Dance of the Universe." Visitors will also be able to exercise on futuristic equipment.
- **ZooDNA**, where visitors are introduced to tomorrow's genetic possibilities in a "zoo without cages" and in a laboratory where new species of plants and animals are created.
- **The Fuzzy Freeway**, a ride through the various attractions of Astropark. "The Freeway is the place where intellect, logic, reason, emotion, and intuition all intersect to produce a powerful Fuzzy Logic zone producing bursts of creativity that no single sector could produce alone," according to the designers.
- **Good Vibrations**, including a sound garden, where human motion stimulates a spontaneous symphony, and the Anechoic Chamber, which eliminates all outside sounds so individuals can hear the soft, rhythmic sounds of their own bodies.
- **Young Astronauts Club Teleport**, a communications and education centre overlooking the Great

Pavilion and giving YAC members instant on-line access to global communications.

The Hyperion Project creators envision their design as a model for a future, self-contained biosphere colony on Mars. "The name Hyperion connotes something at once futuristic and strong," they explain. "Our mythical, future city on Mars appropriately carries this heroic name and grand design. The Hyperion dream combines myth and purpose into a powerful vision of tomorrow. In the magnificent space, human aspirations can soar forward into the horizon of dreams."

Changing the World

THE FAR SIDE
by Gary Larson

On Oct. 23, 1927, three days after its invention,
the first rubber band is tested.

▲▶▼▶▲▼▶▼▶▲▼▶▼▶▲▼▶▼▶▲▼▶▼▶

WOMEN INVENTORS

by Kiley Armstrong

Windshield wipers, bullet-proof vests, and disposable diapers—these are just a few of the diverse products that were invented by women.

Mother is a necessity of invention.

Just ask 71-year-old Anne Macdonald, inventor, mother of five, retired history teacher, and proud chronicler of "feisty women."

Macdonald of Washington, D.C., spent four years doing research that contradicts an ages-old notion that "women just didn't have the brains" to be inventors.

Her book *Feminine Ingenuity* (published by Ballantine Books) celebrates women's creations: windshield wipers, disposable diapers, fire escapes, street sweepers, and bulletproof fibres, to name a few.

"It was intoxicating," Macdonald said of her findings.

This daughter of feminist parents—her mother was one of the first women engineers—attended and taught at girls' schools where equality was "in the water you drink."

At age 17, she invented a knitting device. When she decided to patent it at age 62, her attorney asked condescendingly, "And what is your little idea?"

She won the patent, sans attorney, and successfully marketed the device.

While researching her first book, about knitting, she noticed a reference to women patent-holders in an 1892 magazine called *Domestic Monthly*. Intrigued, she pored over files at the U.S. Patent Office, Library of Congress, and National Archives.

She interviewed Nobel Prize winner Gertrude Elion, the first woman inducted into the Inventor's Hall of Fame. "When she tried to get a job working in a lab, they told her she was too attractive," said Macdonald. But Elion got work during World War II and invented anti-leukemia drugs.

Macdonald also examined the letters of earlier, frustrated inventors, who were "harassing the daylights out of the (patent) examiners—saying, 'You're just slow because you're a man! You don't understand how potato boilers would help women because you don't have to sit in a steamy kitchen!'

"I'm sitting there and thinking, 'Go for it, Mamie!' " said Macdonald. "I see their gutsiness, their stick-to-itiveness, their pride."

An 1876 exposition in Philadelphia featured women's inventions, but the bulk were "homemaker things: a fancy darner that sold like hot-cakes; Mrs. Potts' Iron—there wouldn't have been a 'Mr. Potts' Iron,' " Macdonald noted dryly.

"Many of the feminists were upset that so much of the stuff shown was domestically oriented. They thought housework was a badge of slavery to women."

The years that followed brought a plethora of inventions:

- Mary Anderson, sympathetic to a motorman wiping snow from a New York streetcar, invented a forerunner to modern windshield wipers in 1903.
- Laundry-weary Marion Donovan fashioned shower curtains and absorbent material into the first disposable diaper in 1951. She later sold her company for $1 million.
- Anna Connelly stretched a bridge—complete with guard rails and a bell—between rooftops. She was one of nine women between 1878 and 1888 who invented various fire escapes and alarms.
- Cynthia Westover, secretary to New York's commissioner of street cleaning, invented a street-sweeping device in 1892 that speeded the work of humans and protected horses from back-breaking loads.
- Chemist Stephanie Kwolek invented a solution in 1965 that led to Kevlar: the fibre used for tires, airplanes, space vehicles, boats, and bulletproof vests.

- Actress Hedy Lamarr, after her marriage to a munitions dealer, co-invented a wartime radio signalling device. It was adapted by a manufacturer after her patent expired.

"It's terribly important for young girls to be encouraged," said Macdonald.

And her battle is likely to continue.

During a promotion tour, a male radio interviewer informed Macdonald that inventiveness "was a question of testosterone[1]."

"I almost fell through the microphone," said Macdonald.

1. **testosterone:** a male hormone

▲▼▶▼▲▼▶▼▲▼▶▼▲▼▶▼▲▶

CANADA FIRSTS

by Ralph Nader

Ralph Nader, the famous consumer advocate, is a fan of Canadian ingenuity. He even wrote a book, called *Canada Firsts*, celebrating Canadian achievements.

The Snowmobile

In 1922, mechanic Joseph-Armand Bombardier invented the first "snowmobile" (autoneige in French), a propeller-driven sled. The initial design was for a one- or two-person snowmobile, which he called the "Husky," but it would be some time yet before the machines began to replace sled dogs. The Mounties resisted the change until 1969, when the expense of keeping dogs was cited as a factor in the switch.

Several years passed before Bombardier's ideas would be incorporated into the modern machine. The first commercially successful snowmobile was the B-7 Bombardier Snowmobile, built by Bombardier and patented on June 29, 1937.

Little more than a car placed upon skis and half-tracks, the machine seated seven and cost around 7500 dollars. He sold 50 models and they were used as buses and for medical transport in the winter.

The heavy weight of most engines prevented further development of Bombardier's idea until 1958, when he found a light two-stroke, single-cylinder engine in Austria that weighed less than 14 kg. The following winter, on November 9, 1959, the first 250 snowmobiles, which Bombardier intended to call "Ski-dogs," but changed the name to "Ski-doos" rolled off the assembly line at Bombardier's plant at Valcourt, Quebec (near Sherbrooke). In 1971, 226 000 Ski-doos were sold in North America.

The fascination with travelling over snow must have run in the Bombardier family: the first Canadian to reach the North Pole by snowmobile was Jean-Luc Bombardier, a nephew of Joseph Bombardier.

IMAX Film Format

In 1949, Colin Low of Cardston, Alberta, joined the National Film Board (NFB) as a graphic artist. By 1959, he had become involved in filmmaking, trained in this new profession by NFB's master animator, Norman McLaren. In the 1960s, Low helped develop revolutionary film formats in the film industry. One of the projects he was involved in was co-directing the spectacular film *Labyrinth* for the 1967 World Exposition in Montreal. *Labyrinth* used both 70-mm and 35-mm film and was projected on several screens, giving the viewer 10 times the information that normal films do. This film was the precursor to the IMAX and OMNIMAX film formats now used in many amusement-park films and documentaries.

In 1968, Grahame Ferguson, Roman Kroitor, and Robert Kerr invented IMAX ("I" for eye plus "MAX" for maximum—the most the eye can see) motion pictures for the Canadian pavilion at Expo 1970 in Osaka, Japan. The first permanent IMAX theatre, Cinesphere, opened that same year at Ontario Place, on Toronto's waterfront. There are now 10 permanent IMAX theatres in the United States, and 55 others in 14 countries around the world. IMAX uses 70-mm film turned sideways, the largest film frame in history, producing six-storey-high projected images. The films are produced by IMAX Systems Corporation of Toronto, and the specially designed IMAX film projectors are built in Oakville, Ontario. Two of the documentaries filmed with IMAX cameras are *Hail Columbia* and *The Dream is Alive*, both of which captured the launch of the NASA Space Shuttle. Other IMAX films include *Transitions*, made for Expo 1986 in Vancouver, B.C., and *The Heart*, both of which were directed by Colin Low and Tony Ianzelo, with cameraman Ernest McNabb.

IMAX Systems Corporation unveiled its

Solido system in 1990. The system involves three-dimensional images which are projected upon a dome-shaped screen 24 m in diameter that is tilted to fill the audience's full field of vision from side to side and above. The experience of watching films in such a theatre has been likened to looking and moving around in real visual space. As Roman Kroitor stated at the screening of a demonstration film, "When you're experiencing a Solido film, where the dome-like screen occupies all of the retina, it feels very different. The more of your retina that's occupied by the image, the more powerful it is psychologically."

STOL (Short Take-off and Landing Aircraft)

Since the 1940s, Canada has been a pioneer in the manufacturing of low-speed, low-flying aircraft (at altitudes of 20 000 to 30 000 feet [6096 m to 9144 m]) capable of landing on short runways (less than 610 m). STOL aircraft are also much quieter than standard designs. The Beaver, the first true STOL aircraft, was delivered for use by de Havilland Canada in 1948. The Turbo-Beaver, Otter and Twin Otter, Caribou, and Buffalo were also designed with STOL characteristics by de Havilland Canada. Developed for use by bush pilots and the military, it was realized in the 1960s that STOL aircraft could also reduce the time and energy spent getting to and from airports by making downtown airports feasible. In a demonstration of the assets of STOL aircraft in 1966 that came to be known as Metro 66, 440 takeoffs and landings were held at eight different downtown Manhattan sites without any damage to aircraft, personnel, or property. The largest contingent of the more than 40 aircraft involved was provided by de Havilland Canada. Studies began in both the U.S. and Canada in 1967 to link cities along STOL routes.

In 1971, the Canadian Department of Transport authorized the development of a STOL commuter service for demonstration. By 1973, the U.S. Civil Aeronautics Board was forced to terminate its plans for a Northeast Corridor STOL project, and Canada decided to proceed on its own. A STOL service called Airtransit, operated as a subsidiary of Air Canada and flying between Ottawa and Montreal, began service on July 24, 1974, using six specially modified de Havilland Twin Otters. It was an immediate success and cut downtown-to-downtown air travel time from two hours to one hour and 25 minutes.

Today the Dash-7, a commuter or regional airliner, is in use in over 107 countries around the world, the largest user being the United States. Since 1974, half-hourly flights between Ottawa and Montreal have been featured by Airtransit, and the planes used are 12-passenger de Havilland Canada Twin Otters.

HOW TO BUILD A BETTER POTATO CHIP

by Maureen Murray

Welcome to the world of biotechnology, where scientists change the characteristics of plants and animals through genetic engineering.

Imagine a vine-ripened tomato that doesn't soften or bruise, potatoes that give insects ulcers, and cows that produce more milk.

Welcome to the brave new world of biotechnology, a process that uses gene technology to alter or improve certain characteristics in plants and animals.

An array of works in progress were discussed at a 1993 supermarket and grocery products convention in Chicago. Among them:

• Scientists are experimenting to create a potato that has more starch, which would make a better potato chip. They are also trying to remove a potentially cancer-causing agent found naturally in potatoes.

- There are efforts to make tomatoes resistant to frost by introducing a gene from the flounder. Only the flounder's cold-resisting characteristics would be passed on. Salads will not smell of fish.
- Scientists are attempting to introduce a gene into potatoes and squash which would repel predator insects.

"Just the term biotechnology often is all it takes to conjure up visions of bizarre new life forms such as killer plants or some grotesque animal," said Lester Crawford, a food scientist with the National Food Processors Association, in Washington, D.C.

But, biotechnology holds the key to improving food quality and increasing efficiency in farming, he added. "The most important benefit of biotechnology is a bountiful and safe food supply for the future."

He conceded that foods derived through this process will have trouble gaining acceptance from the general public. Consumers have always been sceptical of attempts to tamper with something as basic as the food they eat. A hundred years ago, when dairies first began pasteurizing milk, the public reacted with outrage, claiming "this process changes the property of food and will put our food supply in peril," Crawford said.

But, he noted, consumers might not be aware of biotechnology-derived foods already on the market.

Canola oil, for instance, naturally contains a substance called erucic acid, which makes it unsuitable for human consumption. Using biotechnology, the substance has been removed, with the result that canola oil can be found on most grocery shelves.

Cheese is traditionally made with an enzyme obtained from the stomach of a calf, but biotechnology has developed a process where bacteria produce the same enzyme and in much larger quantities. About 40 percent of the cheese sold in Canada and the United States is made using this engineered enzyme.

"For any vegetable currently consumed today, there are experiments under way to make it better or more nutritionally sound," Crawford said. Vegetables made more resistant to pests would need less pesticides and be safer to the environment, he added.

While the public has mixed feelings about biotechnology and food, consumers seem to be most concerned about mixing animal and plant genes. "The insertion of animal genes into plants raises significant emotional and ethical concerns," said Thomas Hoban, a psychology professor at North Carolina State University.

The university recently completed a major survey on consumer attitudes toward bioengineered foods. Consumers are generally more accepting of genes from one variety of plants being crossed with another. About 70 percent of the 1200 respondents in the survey said they supported the use of biotechnology in agriculture and food production, Hoban stated.

Selling products with spliced animal and plant genes is somewhat down the road, Crawford said. "It will be another 30 years before you will see a supermarket full of products bioengineered in this way."

But a bioengineered tomato derived from splicing genes from a wild species of the fruit may be available at a produce department near you in the not-too-distant future. This tomato has a 25 percent longer shelf life and is resistant to bruising.

TECHNO-MARVELS

by Malcolm Abrams and Harriet Bernstein

It's a miracle! Now you can ski uphill and walk on water. Some inventors just want to have fun.

Uphill Skiing

How many times have you skied down a challenging slope, only to realize the bigger challenge of getting back up to the top for another run? John Stanford and Phil Huff decided to use their parachuting and skiing experience to design a product that would solve that dilemma. The result is a lightweight parachute powerful enough to propel skiers up steep slopes, yet small enough to be easily packed away for the trip back down.

Coming up with a prototype was fairly simple—since Stanford's company manufactures parachutes—but testing it was downright thrilling. "We realized that uphill skiing was more fun than skiing downhill," says Stanford. So once they received a patent, the sport of "upskiing" was born.

The parachute can be used on snow-covered lakes or steep mountain slopes, in winds as low as 11 or 12 km/h. Stanford and Huff have upskied in 80-km/h winds, but describe the experience as "terrifying" and "dangerous" and strongly advise against it.

Like sailing, upskiing is a wind sport. After putting on your skis and strapping yourself into the harness, lift part of the canopy (with the help of control lines) so that it fills with wind. Then just lean back and go. On a modest slope with firm snow and average winds of about 15 to 25 km/h, you can move at twice the speed of the wind. To get yourself up a very steep slope, however, you'll need stronger winds of about 29 to 32 km/h.

A control centre attached to the harness allows you to increase or decrease your speed and, in the case of an emergency, release yourself from the equipment. The parachute itself is 8.5 m in diameter, and the whole system weighs a mere 6 kg and folds up to the size of a backpack.

Water Walkers

To "walk on water" has long been synonymous with doing the impossible. Which may have been the motivation for Ben Watson's 15-year quest to do just that—develop an apparatus so that humans can stride across lakes and rivers.

Well, Mr. Watson of Renton, Washington, has created water walkers, tested them, obtained a patent, and has sought a manufacturer. He still takes walks on a local lake at about 8 km/h.

His invention looks a bit like a pair of bulky skis (1.8 m long and 23 cm wide), except that each walker is 23 cm deep and the foot is inserted inside a foot well instead of a boot. Each of the walkers, or floats—to be more accurate—is equipped with eight flaps on the underside to propel motion forward. The walkers are made of foam and fibreglass, and a pair weighs 13.5 kg. They can support up to 135 kg. The actual motion one uses to make headway is more like skating or cross-country skiing than walking, as the floats never leave the water.

Now, the big question is, Who wants or needs

to "walk" on water enough to spend $850? Mr. Watson claims his water walkers are stable enough for fishing or hunting and provide a challenging aerobic off-season workout for skiers and other athletes. And one has to admit there would be head-turning excitement about walking across the lake on a bright summer day.

And, "for more fun and excitement," Mr. Watson has also devised sails to convert your walkers into a little catamaran[1].

1. **catamaran:** a sailboat with two hulls, or floaters

DRIVING INTO THE FUTURE

adapted from an article by Patricia Orwen

Smart cars that drive themselves? Sounds amazing, but they may be just around the corner. And they could be absolutely necessary on our overcrowded highways.

Here's what a smart car will soon be able to do: The intelligent vehicle of the future will drive itself more carefully, more systematically, and with less fuel waste and pollution than any human being does today.

It would "see" and follow pavement markings with its videocamera eyes. Its radar would alert its microprocessor brain to obstacles. So if, for example, a cat were to dart suddenly in front of a car, the radar would alert the vehicle's brain, and the brain would activate the motor to slam on the brakes.

If you put the intelligent car on an intelligent highway, it would join a long line-up of cars all taking their orders from a computer at a large traffic monitoring centre. Sound like science fiction? It's not. Smart cars and smart highways are expected to operate in traffic-congested areas around Los Angeles by early in the 21st century. Japan and parts of Europe have plans in motion for a similar system around the same time, and it's quite possible that Canadians will follow suit.

This dream of the automated highway and driverless car will be born out of necessity. "The more cars you've got, the more problems you run into, so the more you need a system—a variety of technologies [to handle traffic]," says Phil Masters, a supervisor with the Ontario Ministry of Transportation.

He should know. Ontario's Highway 401 is the second most traffic-choked highway in North America (after California's Santa Monica Freeway). The part of the 401 that goes through Toronto carries 300 000 cars daily, with motorists averaging 150 000 hours of delay per month. Add a major accident on top of that, and the extra time and cost become astronomical.

Ontario already uses computerized information systems to alert drivers to delays caused by accidents and roadworks. Drivers can then choose alternate routes. The results have been good: cars burn less fuel, cause less pollution, and have fewer accidents.

The evolution of "thinking cars" started in the mid-1980s, with the development of anti-lock brakes and traction control. Both of these features counteract the driver's actions in order to keep the vehicle under control. Car buyers will soon see more such innovations.

The Ford Motor Co. is developing a collision avoidance system, the "all-weather, night-vision sensor." The system uses a radar device that scans both near and faraway objects. "If you were in the fog and couldn't see, the sensor would sound an alarm to warn you about something ahead of you," says Eduardo Peralta, Ford's program manager for advanced concepts. Ultimately, says Peralta, the system will probably be modified so that if the driver doesn't take the necessary action, the car will do it by itself.

Mercedes-Benz researchers in Germany are testing a similar feature called "seeing-eye" cruise control. This system uses infra-red sensors to enable the car to maintain a safe following distance. Should the driver in front of you suddenly slow down, your car will automatically slow down as well.

Toyota, meanwhile, has developed an alarm system aimed at detecting driver fatigue. Using wristbands to monitor the driver's pulse, the car sounds an alarm when the heartbeat slows too much. If the driver remains inattentive, the computer orders the driver's seat to vibrate. If the driver continues to doze, even with these warnings, the vehicle will stop.

Ultimately, some car makers say, the driver will have less and less to do until finally the car will, at least in some circumstances, drive itself. Wow! This gives a whole new meaning to that old word, "*auto*mobile."

▲▼►▼►▲▼►▼►▲▼►▼►▲▼►▼►►

MIRACLE AT KITTY HAWK

by Tom Stacey

By February of 1903, the Wright brothers,
Wilbur and Orville, had already made more
than 1000 glider flights. This fascinating story
tells how they added a heavy gasoline engine
to their flying machine and still got it off the
ground.

Wanted: A Glider with Thrust

When the weather permitted, the Wrights flew
their glider enthusiastically. They gradually
became expert glider pilots, building the skills they
would need later when they attempted powered
flight. After they made more than 1000 glider
flights during a period of three years, the Wrights
felt certain that they could build a powered
machine that would fly. With their innovative
wing-warping idea, they had solved the problem of
lift. Now they had only to figure out how to provide the necessary thrust to power their glider
through the air.

Beginning in February 1903, the Wrights

began gathering the best materials, including wood and fabric, they could find for constructing a powered flying machine, the *Wright Flyer I*. They decided that the thrust would be provided by propellers, powered by a 12-horsepower gasoline engine. The engine was designed by the Wrights and built by Charlie Taylor, a skilful mechanic they had hired to work in the bicycle shop. Taylor's engine would use a chain-and-sprocket design, similar to the bicycle design that the Wrights were so familiar with.

The brothers had debated for many hours what the proper design of the propellers should be. Exactly what would be the perfect shape? How should the propellers be mounted? How fast should they spin and in what direction? These were difficult questions that no one had ever answered.

The Wrights had a peculiar way of debating such questions. They often discussed matters of aerodynamics after dinner in their house in Dayton, Ohio. From the kitchen, their sister Katherine could hear them arguing in the living room. Their voices would grow louder as Wilbur and Orville voiced their disagreements on some point of aircraft design. But often a funny thing happened.

After arguing back and forth for a while, the brothers would suddenly realize that they had switched positions. Wilbur would be supporting Orville's original point, and vice versa. They found that with mutual respect and a lack of stubbornness, they often arrived at the same point of view. This method also allowed each brother to see both sides of a problem and gain a larger perspective on questions of aircraft design. It was almost as if the problem of flight were too difficult for one person alone.

Stormy Weather

Working as a team through the summer months, the Wrights were ready to return to Kitty Hawk, North Carolina, by late September 1903. They brought with them the *Wright Flyer I*. They were eager to try the new craft out. But soon after they arrived, a powerful storm hit, buffeting their wooden cabin with winds up to 110 km/h. Fifteen centimetres of rainwater covered the floor by the time the storm stopped four days later. They cleaned up the camp, thankful that the unassembled pieces of the flyer survived the storm undamaged.

They eagerly put the flyer together. When they started up the motor, however, it backfired violently, causing the entire plane to shake. The engine worked, but it turned the chain and sprockets unevenly. The sprockets were attached to tubular shafts that would spin the propellers. As they began spinning with the sprockets, one of these propeller shafts became badly twisted. The plane would not fly like this. The Wrights knew there was only one thing to do: send it back to Ohio. In Dayton, their helper Charlie Taylor would reinforce both propeller shafts. This work would cause a lengthy delay, but it had to be done. By mid-November, the propeller shafts had been strengthened and were ready to put back on the flyer.

The brothers were stalled again, however. Kitty Hawk was struck with more bad weather through Thanksgiving, including some snow. The weather finally cleared near the end of the month, and the Wrights brought the flyer back out. But while running the engine, they noticed another problem with the tubular propeller shafts. This time one had cracked. It seemed that their hoped-for flight would not happen by Christmas.

Again the Wrights made a quick decision. Orville travelled back to Dayton and had Charlie Taylor construct solid propeller shafts of high-grade steel. Taylor worked quickly, and Orville was back at Kitty Hawk by December 9. Three days later, the improved shafts were in place.

With their mechanical troubles finally solved, the Wrights were eager to try the flyer out the next day. But they needed a certain amount of wind to produce the required lift, and it was just not windy enough. Although the wind picked up the following day, it was Sunday and the Wrights never worked on Sunday. On Monday, December 14, the wind had died down again. But Wilbur and Orville were determined to make a flight, with or without the wind. They had promised that they would be home for Christmas, and time was running out.

Before attempting their first flight, they ran up a flag on a makeshift flagpole to alert the local lifeguards and people on the beach. Five men, two boys, and a dog came to watch. Wilbur won the coin toss to see who would go first. He climbed on, lying down on his stomach. Orville spun the propellers to get the engine started. He then ran alongside, guiding the machine along the 18-m wooden runner they had placed in the sand. Orville also carried a stopwatch, which he started as soon as the flyer left the ground.

At the controls, Wilbur turned the rudder as soon as the plane left the track. The plane immediately went upward, then came down on its tail, damaging one of the wings. It had travelled only about 30 m, and the flight was over in less than four seconds. It was not really a flight, but the brothers were still very excited. They had learned that their design was fundamentally correct, and success could not be far off.

Breakthrough: "How Exciting to Fly!"

The damage to the flyer was slight, and the brothers immediately set about fixing it. A few days later, Orville got his chance to fly. At 10:35 a.m. on December 17, 1903, he climbed onto the flyer. With Wilbur steadying the wing, Orville steered the flyer down the track and took off into the air. His flight lasted a little longer than Wilbur's, a total of about 12 seconds, but he landed the flyer without damage

after he went about 37 m. As soon as it was over, they brought the plane back to the launching pad and tried again. In the next 90 minutes, they made three more flights, with the brothers alternating at the controls. Wilbur made the final flight of the day, covering 260 m and staying up for 59 seconds.

The Wright brothers dramatically changed the course of human history that day. It was the first time a heavier-than-air vehicle had left the earth under its own power, accomplished controlled and sustained flight, then landed at a spot as high as the spot it had taken off from. A new age had begun, although no one realized it. "At the time we flew our first power plane we were not thinking of any practical uses at all," Orville said years later. "We just wanted to show that it was possible to fly.". . .

More so than anyone before them, the Wrights had looked at human flight as a scientific problem to be solved. First, they learned all they could about the subject. Then, they persisted when things became difficult. Their ultimate success was not an accidental discovery or the result of luck. It took several years of rigorous thinking and patient, hard work. Perhaps most of all, it required faith in a dream.

Someone once asked Orville at what point he got the biggest kick out of flying: was it when that first flight left the ground? "No," he said. "I got more thrill out of flying before I had ever been in the air at all—while lying in bed thinking how exciting it would be to fly."

▲ ▼ ▶ ▼▲ ▼ ▶ ▼▲ ▼▶ ▼▲ ▼ ▶ ▼▲ ▶ ▶

THE FLYING MACHINE

by Ray Bradbury

The Emperor understood a great deal about inventions and how they change the world. But were his imperial actions wise or foolish?

In the year A.D. 400, the Emperor Yuan held his throne by the Great Wall of China, and the land was green with rain, readying itself toward the harvest, at peace, the people in his dominion neither too happy nor too sad.

Early on the morning of the first day of the first week of the second month of the new year, the Emperor Yuan was sipping tea and fanning himself against a warm breeze when a servant ran across the scarlet and blue garden tiles, calling, "Oh, Emperor, Emperor, a miracle!"

"Yes," said the Emperor, "the air *is* sweet this morning."

"No, no, a miracle!" said the servant, bowing quickly.

"And this tea is good in my mouth, surely that is a miracle."

"No, no, Your Excellency."

"Let me guess then—the sun has risen and a new day is upon us. Or the sea is blue. *That* now is the finest of all miracles."

"Excellency, a man is flying!"

"What?" The Emperor stopped his fan.

"I saw him in the air, a man flying with wings. I heard a voice call out of the sky, and when I looked up, there he was, a dragon in the heavens with a man in its mouth, a dragon of paper and bamboo, coloured like the sun and the grass."

"It is early," said the Emperor, "and you have just wakened from a dream."

"It is early, but I have seen what I have seen! Come, and you will see it too."

"Sit down with me here," said the Emperor. "Drink some tea. It must be a strange thing, if it is true, to see a man fly. You must have time to think of it, even as I must have time to prepare myself for the sight."

They drank tea.

"Please," said the servant at last, "or he will be gone."

The Emperor rose thoughtfully. "Now you may show me what you have seen."

They walked in to a garden, across a meadow of grass, over a small bridge, through a grove of trees, and up a tiny hill.

"There!" said the servant.

The Emperor looked into the sky.

And in the sky, laughing so high that you

could hardly hear him laugh, was a man; and the man was clothed in bright papers and reeds to make wings and a beautiful yellow tail, and he was soaring all about like the largest bird in a universe of birds, like a new dragon in a land of ancient dragons.

The man called down to them from high in the cool winds of morning, "I fly, I fly!"

The servant waved to him, "Yes, yes!"

The Emperor Yuan did not move. Instead he looked at the Great Wall of China now taking shape out of the farthest mist in the green hills, that splendid snake of stones which writhed with majesty across the entire land. That wonderful wall which had protected them for a timeless time from enemy hordes and preserved peace for years without number. He saw the town, nestled to itself by a river and a road and a hill, beginning to waken.

"Tell me," he said to his servant, "has anyone else seen this flying man?"

"I am the only one, Excellency," said the servant, smiling at the sky, waving.

The Emperor watched the heavens another minute and then said, "Call him down to me."

"Ho, come down, come down! The Emperor wishes to see you!" called the servant, hands cupped to his shouting mouth.

The Emperor glanced in all directions while the flying man soared down the morning wind. He saw a farmer, early in his fields, watching the sky, and he noted where the farmer stood.

The flying man alit with a rustle of paper and a creak of bamboo reeds. He came proudly to the Emperor, clumsy in his rig, at last bowing before the old man.

"What have you done?" demanded the Emperor.

"I have flown in the sky, Your Excellency," replied the man.

"What *have* you done?" said the Emperor again.

"I have just told you!" cried the flier.

"You have told me nothing at all." The Emperor reached out a thin hand to touch the pretty paper and the birdlike keel of the apparatus. It smelled cool, of the wind.

"Is it not beautiful, Excellency?"

"Yes, too beautiful."

"It is the only one in the world!" smiled the man. "And I am the inventor."

"The *only* one in the world?"

"I swear it!"

"Who else knows of this?"

"No one. Not even my wife, who would think me mad with the sun. She thought I was making a kite. I rose in the night and walked to the cliffs far away. And when the morning breezes blew and the sun rose, I gathered my courage, Excellency, and leaped from the cliff. I flew! But my wife does not know of it."

"Well for her, then," said the Emperor. "Come along."

They walked back to the great house. The sun was full in the sky now, and the smell of the grass was refreshing. The Emperor, the servant, and the flier paused within the huge garden.

The Emperor clapped his hands. "Ho, guards!"

The guards came running.

"Hold this man."

The guards seized the flier.

"Call the executioner," said the Emperor.

"What's this!" cried the flier, bewildered. "What have I done?" He began to weep, so that the beautiful paper apparatus rustled.

"Here is the man who has made a certain machine," said the Emperor, "and yet asks us what he has created. He does not know himself. It is only necessary that he create, without knowing why he has done so, or what this thing will do."

The executioner came running with a sharp silver axe. He stood with his naked, large-muscled arms ready, his face covered with a serene white mask.

"One moment," said the Emperor. He turned to a nearby table upon which sat a machine that he himself had created. The Emperor took a tiny golden key from his own neck. He fitted this key to the tiny, delicate machine and wound it up. Then he set the machine going.

The machine was a garden of metal and jewels. Set in motion, birds sang in tiny metal trees, wolves walked through miniature forests, and tiny people ran in and out of sun and shadow, fanning themselves with miniature fans, listening to the tiny emerald birds, and standing by impossibly small but tinkling fountains.

"Is *it* not beautiful?" said the Emperor. "If you asked me what I have done here, I could answer you well. I have made birds sing, I have made forests murmur, I have set people to walking in this woodland, enjoying the leaves and shadows and songs. That is what I have done."

"But, oh, Emperor!" pleaded the flier, on his knees, the tears pouring down his face. "I have done a similar thing! I have found beauty. I have flown on the morning wind. I have looked down on all the sleeping houses and gardens. I have smelled the sea and even seen it, beyond the hills, from my high place. And I have soared like a bird; oh, I cannot say how beautiful it is up there, in the sky, with the wind about me, the wind blowing me here like a feather, there like a fan, the way the sky smells in the morning! And how free one feels! *That* is beautiful, Emperor, that is beautiful too!"

"Yes," said the Emperor sadly, "I know it must be true. For I felt my heart move with you in the air and I wondered: What is it like? How does it feel? How do the distant pools look from so high? And how my houses and servants? Like ants? And how the distant towns not yet awake?"

"Then spare me!"

"But there are times," said the Emperor, more sadly still, "when one must lose a little beauty if one is to keep what little beauty one already has. I do not fear you, yourself, but I fear another man."

"What man?"

"Some other man who, seeing you, will build a thing of bright papers and bamboo like this. But the other man will have an evil face and an evil heart, and the beauty will be gone. It is this man I fear."

"Why? Why?"

"Who is to say that someday just such a man, in just such an apparatus of paper and reed, might not fly in the sky and drop huge stones upon the Great Wall of China?" said the Emperor.

No one moved or said a word.

"Off with his head," said the Emperor.

The executioner whirled his silver axe.

"Burn the kite and the inventor's body and bury their ashes together," said the Emperor.

The servants retreated to obey.

The Emperor turned to his hand-servant, who had seen the man flying. "Hold your tongue. It was all a dream, a most sorrowful and beautiful dream. And that farmer in the distant field who also saw, tell him it would pay him to consider it only a vision. If ever the word passes around, you and the farmer die within the hour."

"You are merciful, Emperor."

"No, not merciful," said the old man. Beyond the garden wall he saw the guards burning the beautiful machine of paper and reeds that smelled of the morning wind. He saw the dark smoke climb into the sky. "No, only very much bewildered and afraid." He saw the guards digging a tiny pit wherein to bury the ashes. "What is the life of one man against those of a million others? I must take solace from that thought."

He took the key from its chain about his neck and once more wound up the beautiful miniature garden. He stood looking out across the land at the Great Wall, the peaceful town, the green fields, the rivers and streams. He sighed. The tiny garden whirred its hidden and delicate machinery and set itself in motion; tiny people walked in forests, tiny foxes loped through sun-speckled glades in beautiful shining pelts, and among the tiny trees flew little bits of high song and bright blue and yellow colour, flying, flying, flying in that small sky.

"Oh," said the Emperor, closing his eyes, "look at the birds, look at the birds!"

5

Roll on,
Robots!

Portrait of a Machine

What nudity is beautiful as this
Obedient monster purring at its toil;
These naked iron muscles dripping oil
And the sure-fingered rods that never miss.
This long and shining flank of metal is
Magic that greasy labor cannot spoil;
While this vast engine that could rend the soil
Conceals its fury with a gentle hiss.

It does not vent its loathing, does not turn
Upon its makers with destroying hate.
It bears a deeper malice; throbs to earn
Its master's bread and lives to see this great
Lord of the earth, who rules but cannot learn,
Become the slave of what his slaves create.

Louis Untermeyer

▲▼▶▼▶▲▼▶▼▶▲▼▶▼▶▲▼▶▼▶▲▼▶▼▶▶

ROBOT CONTROL

by Rhys Lewis

Once only imagined in science fiction, robots have assembled an important place for themselves in the working world.

A robot is a machine that can be instructed to perform tasks which would otherwise have to be done by people. Most robots have computer "brains," which can be programmed by an operator to carry out specific tasks. The latest robots are equipped with sensors which enable them to see, hear, touch, and even smell their surroundings. They have an electronic brain sophisticated enough for them to decide how to act in response to this data. Hundreds of thousands of robots are now at work, and in future, with advances in computer technology, robots will have even more complex "brains," which will enable them to learn from experience.

Robot Origins

Before 1920 no one had ever heard of robots. The word was coined by the Czechoslovakian playwright Karel Capek, from the Czech words "robota," meaning slave labour, and "robotnik," meaning slave. In his play *R.U.R.* (Rossum's Universal Robots), robots were intelligent, hardworking, human-shaped machines which rebelled against their human masters and killed them when the robots were used for making war. Since then, robots have featured in many books and films, sometimes as good and helpful, but often as evil and destructive.

It is their ability to be instructed which distinguishes robots from machines that imitate human actions by mechanical means. Two thousand years ago, the Greeks and Romans could make statues that moved by hydraulic power. In the 1700s and 1800s there were life-size clockwork models which could even write and play musical instruments. However, these were only mechanical toys, and the correct name for them is "automata."

In the 1950s machines came into use that could operate automatically, following instructions on punched paper tape or magnetic tape. These and later computer-controlled machines ushered in increasing automation in industry, bringing about a second industrial revolution. In the last few decades tireless computer-controlled robots have taken their place alongside human beings on factory production lines.

Designing Robots

The human body is like a superbly engineered machine controlled by an incredibly complex computer: the brain. It will be a long, long time before robots are capable of responding to their environment, learning from experience, and thinking creatively with the dexterity of the human brain. Arms and hands are the human limbs whose functions are most often duplicated by robots. To have the necessary freedom of movement, robot arms, like human ones, have joints capable of moving independently of one another, up and down, side to side, and in and out. Robot arms can do some things better than ours, such as picking up heavy objects with ease, or rotating their "wrists" in a full circle.

The robot must be given a computer "brain" which will assess the difficulty of the task it has been given and decide how to set about achieving it. Suppose that a robot needs to pick up an object from a conveyor belt and place it in a box (a relatively simple task for a human). To do this it needs an electronic "eye" to identify the object; if the object is not lying in the expected position, the "eye" relays this information to the computer, which reacts by instructing the arm to approach the object from a different angle to pick it up.

Many of today's robots can use "feedback." This is the ability to make decisions which depend on information about changing conditions. We have nerves to carry instructions from our brains to our muscles. Robots have electronic cables which transmit instructions from the computer to its motor-driven parts.

It is rare for robots to walk upright on two legs. This is because walking on two legs involves lifting one foot from the ground at every step, and becoming unstable. To give a robot the balance and coordination necessary to walk on two legs would take up a wasteful amount of space in its computer brain. So most mobile robots move on wheels, or on four or six legs.

Spot Facts

- Robotized trains run in the San Francisco Bay area of California. At peak times, 100 driverless trains run at high speed with only a 90-second interval between them.

- The automatic pilot of an aircraft is a typical "invisible" robot. An airplane may be landed by automatic pilot with a chance of failure of one in 10 million.

- In 1988 there were 175 000 industrial robots in Japan. There were less than half that number in the rest of the world.

- The earliest kind of robot was a water-clock, called a clepsydra, which was invented by Ctesibus of Alexandria, Egypt, in about 250 B.C. It recycled its water by means of a siphoning device which ran automatically.

▲ ▶ ▼ ▲ ▼ ▶ ▲ ▼ ▶ ▼ ▲ ▼ ▶ ▼ ▲ ▼ ▶ ▼ ▲ ▶

THE FUN THEY HAD

by Isaac Asimov

Way back in the 1950s, sci-fi writer Isaac Asimov foresaw the day when robots and computers would take over completely from human teachers.

Margie even wrote about it that night in her diary. On the page headed May 17, 2155, she wrote, "Today Tommy found a real book!"

It was a very old book. Margie's grandfather once said that when he was a little boy his grandfather told him there was a time when all stories were printed on paper.

They turned the pages, which were yellow and crinkly, and it was awfully funny to read words that stood still instead of moving the way they were supposed to—on a screen, you know. And then, when they turned back to the page before, it had the same words on it that it had had when they read it the first time.

"Gee," said Tommy, "what a waste. When you're through with the book, you just throw it away, I guess. Our television screen must have had a million books on it and it's good for plenty more. I wouldn't throw it away."

"Same with mine," said Margie. She was eleven and hadn't seen as many telebooks as Tommy had. He was thirteen.

She said, "Where did you find it?"

"In my house." He pointed without looking, because he was busy reading. "In the attic."

"What's it about?"

"School."

Margie was scornful. "School? What's there to write about school? I hate school." Margie always hated school, but now she hated it more than ever. The mechanical teacher had been giving her test after test in geography and she had been doing worse and worse until her mother had shaken her head sorrowfully and sent for the County Inspector.

He was a round little man with a red face and a whole box of tools with dials and wires. He smiled at her and gave her an apple, then took the teacher apart. Margie had hopd he wouldn't know how to put it together again, but he knew how all right and, after an hour or so, there it was again, large and black and ugly with a big screen on which all the lessons were shown and the questions were asked. That wasn't so bad. The part she hated most was the slot where she had to put homework and test papers. She always had to write them out in a punch code they made her learn when she was six years old, and the mechanical teacher calculated the mark in no time.

The inspector had smiled after he was finished and patted her head. He said to her mother, "It's not the little girl's fault, Mrs. Jones. I think the geography sector was geared a little too quick. Those things happen sometimes. I've slowed it up to an average ten-year level. Actually, the overall pattern of her progress is quite satisfactory." And he patted Margie's head again.

Margie was disappointed. She had been hoping they would take the teacher away altogether. They had once taken Tommy's teacher away for nearly a month beause the history sector had blanked out completely.

So she said to Tommy, "Why would anyone write about school?"

Tommy looked at her with very superior eyes. "Because it's not our kind of school, stupid. This is the old kind of school that they had hundreds and hundreds of years ago." He added loftily, pronouncing the word carefully, "Centuries ago."

Margie was hurt. "Well, I don't know what kind of school they had all that time ago." She read the book over his shoulder for a while, then said, "Anyway, they had a teacher."

"Sure they had a teacher, but it wasn't a *regular* teacher. It was a man."

"A man? How could a man be a teacher?"

"Well, he just told the boys and girls things and gave them homework and asked them questions."

"A man isn't smart enough."

"Sure he is. My father knows as much as my teacher."

"He can't. A man can't know as much as a teacher."

"He knows almost as much I betcha."

Margie wasn't prepared to dispute that. She said, "I wouldn't want a strange man in my house to teach me."

Tommy screamed with laughter, "You don't know much, Margie. The teachers didn't live in the house. They had a special building and all the kids went there."

"And all the kids learned the same thing?"

"Sure, if they were the same age."

"But my mother says a teacher has to be adjusted to fit the mind of each boy and girl it teaches and that each kid has to be taught differently."

"Just the same, they didn't do it that way then. If you don't like it, you don't have to read the book."

"I didn't say I didn't like it," Margie said quickly. She wanted to read about those funny schools.

They weren't even half finished when Margie's mother called, "Margie! School!"

Margie looked up. "Not yet, mamma."

"Now," said Mrs. Jones. "And it's probably time for Tommy, too."

Margie said to Tommy, "Can I read the book

some more with you after school?"

"Maybe," he said, nonchalantly. He walked away whistling, the dusty old book tucked beneath his arm.

Margie went into the schoolroom. It was right next to her bedroom, and the mechanical teacher was on and waiting for her. It was always on at the same time every day except Saturday and Sunday, because her mother said little girls learned better if they learned at regular hours.

The screen was lit up, and it said: "Today's arithmetic lesson is on the addition of proper fractions. Please insert yesterday's homework in the proper slot."

Margie did so with a sigh. She was thinking about the old schools they had when her grandfather's grandfather was a little boy. All the kids from the whole neighbourhood came, laughing and shouting in the schoolyard, sitting together in the schoolroom, going home together at the end of the day. They learned the same things so they could help one another on the homework and talk about it.

And the teachers were people....

The mechanical teacher was flashing on the screen: "When we add the fractions 1/2 and 1/4...."

Margie was thinking about how the kids must have loved it in the old days. She was thinking about the fun they had.

ROBOTS AT WORK

As robots become smaller and more flexible, you never know where you might meet one— at a fast-food joint, at the zoo, fighting a fire, or down on the farm!

McRobot

by Bill Lawren

Someday your fast-food burger may be prepared by the fastest of all possible cooks: a piece of computer-controlled machinery. McDonald's, the people who gave us the clamshell grill that cooks hamburgers on both sides simultaneously, has taken another step in that direction.

The company has graced a number of its 12 000 restaurants with fully automated systems that fry your fries or pour your Cokes without the help of *Homo sapiens*. In fact, to get you your soft drink, the crew simply punches your order into the register, and presto! a cup is automatically filled with the appropriate flavour. "All the crew has to do," says McDonald's spokesperson Jane Hulbert, "is ice it."

McDonald's engineering department in Oak Brook, Illinois, developed the system, known as automated restaurant crew helper (ARCH). Restaurants in Colorado, Indiana, Minnesota, and Germany now have ARCH on-line, Hulbert says, and the system is "available to any restaurant in our system. It's up to the individual owner-operator."

Some observers have voiced concern that ARCH may be the harbinger of a fully automated brave new McWorld that deprives teenagers of much-needed jobs. "Absolutely not," Hulbert states. Actually, she says, ARCH gives McDonald's employees "a tremendous opportunity to work with technology. The crew just loves it."

Robotic Eggs to the Rescue

from The Futurist *magazine*

New technologies are coming to the aid of zoologists working to save endangered species. One new tool to rescue bird species is the "robo-egg," an egg-shaped electronic device that a mother bird treats as one of her own eggs. Sensors inside the robo-egg measure temperature and humidity in the nest and how often the mother bird turns the egg. The information is then relayed to researchers.

The robo-egg is already in service at New York's Bronx Zoo, where zoologists are seeking ways to save the highly endangered white-naped crane, reports Smithsonian Institution writer Marc Gretzfelder.

Robotic eggs are tools being used by today's zoo biologists. The aim is to transform zoological parks into more natural homes for the animals, as well as breeding centres for those animals in danger of extinction.

Fire-fighting Robots

from **The Futurist** *magazine*

Exceptionally dangerous fires such as the well-head fires in Kuwait (during the Gulf War in 1991) may increasingly be fought by robots. Remotely operated robotic vehicles will move against fires in explosive atmospheres, radiation areas, highly toxic chemical environments, or other dangerous situations. Robotic firefighters developed at SubSea Offshore, Ltd., in Aberdeen, Scotland, can observe and survey a fire, then quench it with an assortment of tools. Operators controlling the vehicle stay in a sealed, air-conditioned van safely away from the fire.

Robot Farming

from **The Futurist** *magazine*

A robotic harvester that can identify whether melons and other crops are ripe enough to pick may revolutionize farming. By the year 2000, a farmer will be able to purchase such a robot for the price of a new pickup truck, according to agricultural engineer Gaines Miles at Indiana's Purdue University.

Current harvesting machines not only are unable to tell if what they have picked is ripe, but they also often destroy good crops. That's where the Robotic Melon Picker (ROMPER) comes in. The ROMPER is able to transplant, cultivate, and harvest round, or "head," crops such as melons, lettuce, cabbage, and pumpkins.

The ROMPER is smarter than traditional harvesting machines. The device, mounted on a large trailer frame, has two cameras that scan the plants while a fan blows aside leaves to expose hidden produce. A computer then analyzes the images, looking for a round, bright spot and identifying it as the crop to be picked.

The ROMPER then confirms the crop's ripeness by "smell." Ripeness in fruits such as melons can be gauged by ethylene, "a naturally occurring hormone that causes melons to ripen and triggers the release of aromatic volatiles [smells]," according to Meny Benady, a graduate student participating in the Purdue project. Under laboratory conditions, these sensors have gauged ethylene levels and judged ripeness accurately to within a day.

The electronic components for an agricultural robot could cost less than $2000 in the next decade, Miles believes. As the costs of technology drop, more farmers will be able to afford such devices.

"We're still in the early stages of prototype testing," says Miles. "But I believe that by the end of the decade [the 1990s] you'll see at least a few robotic machines in every Indiana county."

▲▼▶▼▶▲▼▶▼▶▲▼▶▼▶▲▼▶▼▶▶

ARTIFICIAL MUSCLE

by Michael Smith

Artificial muscle: it could be the key to a new generation of robots that are "almost" human!

Slipping a virtual-reality helmet over her head and gripping a control stick, the surgeon guides a microscopic probe as it crawls—using its own artificial muscles—through the tiny cavities of a patient's body.

Faced with a 160-kg bedridden woman, a nurse slips a sheath of artificial muscle over his arms and shoulders, adjusts a control and—presto!—he's three times as strong as normal. Turning the patient is child's play.

The common thread in those two scenes is artificial muscle—something that's almost, but not quite, off the drawing board in Ian Hunter's McGill University lab. Hunter, a fellow of the Canadian Institute for Advanced Research, says artificial muscle, using microscopic metal alloy fibres, may "break the deadlock in robotics."

Deadlock?

Well, people have been predicting robots for a long time now. But so far the only widely-used ones are big clunky things that have to be anchored to a factory floor.

"We can build a reasonable brain," Hunter says, "but we can't build decent limbs." The road-block is that the motors that drive robots are just too clumsy. "When we want to make a big machine, we basically have to make a big engine."

Nature, on the other hand, follows a different strategy. Millions of microscopic muscle fibres work together, whether in a flea or a blue whale, to power an animal. "The muscle is all much the same," Hunter says, "but the blue whale has an enormous number of the fibres."

Hunter's group, part of the institute's artificial intelligence program, has already found a way to make microscopic nickel-titanium fibres contract and relax the same way muscle fibres do. "We didn't discover that they contract; that's been known for some time," Hunter says. "What we found was a way to make them contract and relax quickly and at room temperature."

The difficulty is to control the fibres so they work the same way real muscles do. Hunter is working with the institute's theoreticians to solve that problem. "This is new technology and we have to understand it thoroughly in order to make applications," he says.

Tiny Robot Probes for Surgery

Hunter is building a prototype—including the control systems—of the microscopic probe that surgeons could use to scan the tiniest parts of the body.

Doctors already use such probes, called endoscopes, but they have to be thick and stiff enough to be pushed into the body; if they're too flimsy they bend under the pressure. That limits their use. There are places that are simply inaccessible to a probe that's 10 mm across. Hunter's probe, surrounded by its sheath of artificial muscle, would be only half a millimetre across. And it wouldn't have to be pushed. "It would actually worm its way into your body," Hunter says.

Controlling the probe's muscles is a tough problem, but Hunter expects to have it pretty much solved by the mid-1990s. The apparatus that the doctor would use is also tricky, but that's much closer to reality. Essentially, it's a pair of cameras mounted in such a way that they mimic the way your eyes, head, and neck can move. A laser guides the focus, and the two images are combined in a helmet to produce a three-dimensional picture for the surgeon.

The helmet picks up the doctor's head and eye motions and moves the cameras to match. The microscopic probe—when it's ready to go—will respond in much the same way: a doctor will be able to tell the device to look left or right just by turning her head.

Superstrength from Artificial Muscles

Hunter is working on one other main application—the sheath of artificial muscle that he calls "exo-muscle." That's a bit further off, but it's an obvious use of artificial muscle: to make people themselves stronger.

In use, exo-muscle would work rather like power steering, Hunter says. Your own muscles would guide the artificial ones, but they would apply greater force, for a longer time, and without getting tired. "There are many, many applications where it would be nice to have an extra force application capacity," he says. "And we would like to be able to apply that force without getting fatigued."

In the long run, he says, the metal fibres he's working with now probably won't be the basis for generally used artificial muscle, although the control techniques he's developing will continue to be important.

Instead of metal, manufacturers will use fibres made from polymers—molecules made from large numbers of smaller molecules that have been chemically united. The immediate advantage of polymers, Hunter says, is that they will generate more power for less energy than his nickel-titanium fibres.

But, looking into the blue-sky future, Hunter is dreaming of even more advantages, capitalizing on the fact that polymers of various kinds can be used as sensors and as elements in a computer.

The four elements of a living being, he says,

are muscles, sense organs, a brain, and some sort of scaffolding (such as bone) to carry them. It may be possible to "grow" mixed polymers that would combine the muscle function with sensors and computing elements, Hunter thinks.

And if that's the case, "you have all you need to build an artificial organism."

Epilogue:

"BACK" TO THE FUTURE

by Portia Jorgenson

Who's to say which new lifestyle gadgets will become hits and which will bomb? Here are a few "best guesses" that might be coming your way in the near future.

The Moller 400

With heavy congestion on every city's roads and the need people always seem to have for a "new kind of toy" to play with, the "Moller 400" should prove popular. It will offer the appearance of a Corvette and the performance of a rocketship. Taking off vertically, hovering, and capable of 240 km/h, it's a hot ride, for sure!

"Virtual World"

After you have parked the Moller 400 in the garage, it's time to experience faraway places and different times with "Virtual World," a home entertainment centre that is totally out of this world. Never been to Jamaica? Put on your headset, a pair of special goggles, a Virtual World suit and gloves, and enjoy the experience, mon. You'll hear the sound of the surf, feel the heat of the sun, and have a sweeping view of your surroundings. Don't worry, you won't be lonely. A friend wearing the same garb can join you from anywhere around the world. And the fun's not over yet!! You can cast yourself in your favourite role in each new location. Visit the Old West as Billy the Kid. Or, look, feel, and move like an elephant on an African Safari.

Animan

When you get in the mood for lighter and less involving entertainment, count on "Animan," a walking TV with all the personality of a pet. This free-roving machine will follow you from room to room as you do those Saturday morning chores. Groove with the tube, as Animan has had dance lessons. Add drama to action-packed chase scenes by leaning into the curves. Animan by day becomes Aniscout by night, as he protects the home, even sounding an alarm if his security camera detects a burglar. (So much for sneaking in after curfew.)

The Lawn Ranger

If you don't have Animan to help you with the chores, call in the Lawn Ranger. This computer-guided robotic mower will save you from cutting that overgrown green area your parents call the lawn. Designed to replace the push mower, it will steer itself around the yard, mowing any area where the grass is too long. When the job is done, the Lawn Ranger automatically turns itself off.

3-D Outfits

So the work's all done and it's party time! As you thrash your closet looking for something to wear to the most important event of the year, you remember that a new outfit is only 45 seconds away, thanks to your "Ultrasonic 3-D Clothes Computer." You can choose the colour, fabric, details, have body scans, and create the outfit of the future in a matter of seconds. The cloth is made into a three-dimensional shape, cut, and ultrasonically sealed. The best feature is that thanks to the body scan, your personalized duds will fit perfectly. Who needs Calvin Klein?

Mega Ball

If you ever get bored with all these toys, slip out to the amusement park of the future and buy a ticket on "Mega Ball." This one will appeal to those with a strong stomach, who will sit in bumper cars and be launched into a huge pinball machine. Once inside, the human pinballs are bounced through a maze of flashing lights and electronic buzzers. Those with sadistic minds will stand outside the ride and, using the coin-operated push-buttons, send the helpless riders back up the platform for another shot.

As hot as some of these future items sound, there are just as many turkeys on the drawing board, including:

- Gold-flecked toothpaste: Doubt Colgate has much to worry about.
- Commercial stutter: TV ads that are shown twice in a row for impact. "I bored two friends, and they bored two friends, and so on, and so on."
- The mood suit: A bathing suit that reveals the temperature of the torso the suit is covering. Hot bod!

But who's to say what products will finally fly and what won't? Consider what IBM's chairman of the board, Thomas Watson, had to say in 1943: "I think there is a world market for about five computers."

ACKNOWLEDGEMENTS

Permission to reprint copyright material is gratefully acknowledged. Every reasonable effort to trace the copyright holders of materials appearing in this book has been made. Information that will enable the publisher to rectify any error or omission will be welcomed.

Things to Come: Experts Gaze into the Future by Curtis Slepian. Published in the May 1991 issue of *3-2-1 Contact*, a publication of the Children's Television Workshop. Copyright © 1991 Children's Television Workshop (New York, NY). All rights reserved. **Man on Moon** by Stanley Cook © Stanley Cook 1985. From *Spaceways: An Anthology of Space Poetry*, ed. John Foster; published by Oxford University Press. Used with permission. **Spinoffs from Space** by Patricia Barnes-Svarney from "Spinoffs: Technology from the Space Program." *Odyssey*, June 1992; © 1992, Cobblestone Publishing, Inc., 7 School St., Peterborough, NH 03458. Reprinted by permission of the publisher. (Metric conversions and third sentence supplied by Nelson Canada.) **Space Q and A** from "Spinoffs: Technology from the Space Program." *Odyssey*, June 1992; © 1992, Cobblestone Publishing, Inc., 7 School St., Peterborough, NH 03458. Reprinted by permission of the publisher. **Satellites in Space** by Jack R. White reprinted by permission of G.P. Putnam's Sons from *Satellites of Today and Tomorrow*, copyright © 1985 by Jack R. White. **Weightlessness Training** by Gloria Skurzynski. Reprinted with permission of Bradbury Press, an Affiliate of Macmillan, Inc., from *Almost the Real Thing* by Gloria Skurzynski. Copyright © 1981 by Gloria Skurzynski. **The Space Shuttle Disaster** by James McCarter from *The Space Shuttle Disaster* by James McCarter, published in 1988 by Wayland (Publishers) Ltd. Reprinted by permission of Wayland (Publishers) Ltd. **Last Ride** by Andrea Holtslander from *Themes on the Journey* © Nelson Canada, A Division of Thomson Canada Limited, 1989. Reprinted by permission of Andrea Holtslander. **Space Junk** by Judy Donnelly and Sydelle Kramer from *Space Junk: Pollution Beyond the Earth* by Judy Donnelly and Sydelle Kramer, published by Morrow Junior Books. Copyright © 1990 by Wayfarer Press, Inc. **Computer-produced illustration** on page 39 by Teledyne Brown Engineering from *Space Junk: Pollution Beyond the Earth*.

Acknowledgements

Careers in Space from "Space: The Frontier Is Closer Than You Think." *TG Magazine*, December 1992. Reprinted with permission of *TG Magazine*. **Shoe** by Jeff MacNelly reprinted by permission: Tribune Media Services. **What's Your Technotype?** by Carolyn Leitch from *Report on Business*, August 24, 1993. Reprinted by permission of Globe Information Services. **Wearable Computers** from *The Futurist*, September-October 1992. Reproduced with permission from *The Futurist*, published by the World Future Society, 7910 Woodmont Ave., Suite 450, Bethesda, MA 20814. **An Interview with Computer Inventor Steve Wozniak** by Kenneth A. Brown from *Inventors at Work: Interviews with 16 notable American inventors* by Kenneth A. Brown. Published by Tempus Books of Microsoft Press. Copyright © 1988 by Microsoft Press. **Program Loop** by Jill Paton Walsh from *Out of Time: Stories of the Future*, ed. Aiden Chambers, published by The Bodley Head. Copyright © Aidan Chambers 1984. Reprinted by permission of The Bodley Head. **The Far Side** by Gary Larson (page 75) from *Unnatural Selections* by Gary Larson. Copyright © 1991 by Universal Press Syndicate. Reprinted with permission. All rights reserved. **The Bungee Lunge** by Karen McNulty adapted from "Sports Science." *Science World*, February 21, 1992. Copyright © 1992 by Scholastic Inc. Reprinted by permission. **Talking to the Airwaves: Interactive TV** by Lila Gano from *Television: Electronic Pictures* by Lila Gano, a volume in Lucent Books' Encyclopedia of Discovery and Invention series, © 1990 by Lucent Books, San Diego, California. Used with permission. **High-Tech Fashions** by Bernadette Morra adapted from "High-tech fabrics offer comfort in the cold." *The Toronto Star*, January 14, 1993. Reprinted with permission — The Toronto Star Syndicate. **Get Real! The World of Virtual Reality** by Nancy Day excerpted from "Spinoffs: Technology from the Space Program."*Odyssey*, June, 1992. © 1992, Cobblestone Publishing, Inc., 7 School St., Peterborough, NH 03458. Reprinted by permission of the publisher. (Metric conversions supplied by Nelson Canada). **Virtual Ski Training** by Josh Lerman from "Virtue Not to Ski?" *Skiing*, February 1993. Reprinted with permission. **Virtual Cathedral** by Tim Folger from *Discover*, December 1992. © 1992 The Walt Disney Co. Reprinted with permission of *Discover* Magazine. **Designing a City for Mars** from *The Futurist*, March-April 1993. Reproduced with permission from *The Futurist*, published by the

Acknowledgements

World Future Society, 7910 Woodmont Avenue, Suite 450, Bethesda, Maryland 20814, USA. **Illustration of "The Hyperion Project"** (page 93) — Architecture by Michels Bollinger, Inc. Illustration by Peter Bollinger. Reproduced with permission. **The Far Side** by Gary Larson (page 97) from *The Pre-History of the Far Side* by Gary Larson reprinted by permission of Chronicle Features, San Francisco, CA. All rights reserved. **Women Inventors** by Kiley Armstrong from *The Toronto Star*. July 9, 1992. Reprinted by permission of Associated Press. **Canada Firsts** by Ralph Nader from *Canada Firsts* by Ralph Nader. Copyright © 1992 Center for Study of Responsive Law. Published by McClelland & Stewart Inc. **How to Build a Better Potato Chip** by Maureen Murray from *The Toronto Star*, May 11, 1993. Reprinted with permission – The Toronto Star Syndicate. **Mother Goose and Grimm** by Mike Peters, © 1993 Grimmy Inc. Used with permission of Tribune Media Services. **Techno-Marvels** by Malcolm Abrams and Harriet Bernstein from *Future Stuff* by Malcolm Abrams and Harriet Bernstein. Copyright © 1989 by Malcolm Abrams and Harriet Bernstein. Used by permission of Viking Penguin, a division of Penguin Books USA Inc. **Illustration of Uphill Skiing** (page 112) from *Future Stuff* by Malcolm Abrams and Harriet Bernstein. Copyright © 1989 by Viking Penguin. Used by permission of Viking Penguin, a division of Penguin Books USA Inc. **Driving into the Future** adapted from an article by Patricia Orwen in *The Toronto Star*. September 19, 1993. Reprinted with permission – The Toronto Star Syndicate. **Miracle at Kitty Hawk** by Tom Stacey from *Airplanes: The Lure of Flight* by Tom Stacey, a volume in Lucent Books' Encyclopedia of Discovery and Invention series, © 1990 by Lucent Books, San Diego, California. **The Flying Machine** by Ray Bradbury from *The Golden Apples of the Sun* by Ray Bradbury, published by Granada Publishing in Hart-David, MacGibbon Ltd. Reprinted by permission of Don Congdon Associates, Inc. Copyright © 1953, renewed 1981 by Ray Bradbury. **Portrait of a Machine** by Louis Untermeyer from *A Galaxy of Verse* by Louis Untermeyer. Copyright © 1978 by the Estate of Louis Untermeyer. Reprinted with expressed permission by Lawrence S. Untermeyer. **Robot Control** by Rhys Lewis from *The Computer Age* by Rhys Lewis. Published by Wayland (Publishers) Ltd. Copyright © Andromeda Oxford Ltd., 1991. Reprinted with permission. **The Fun They Had** by

Acknowledgements

Isaac Asimov from *Earth Is Room Enough* by Isaac Asimov. Copyright © 1957 by Isaac Asimov. Used by permission of Doubleday, a division of Bantam Doubleday Dell Publishing Group, Inc. **Robots at Work** from *The Futurist*, September-October 1992. Reproduced with permission from *The Futurist*, published by the World Future Society, 7916 Woodmont Ave., Ste. 450, Bethesda, MA 20814. **Artificial Muscle** by Michael Smith from *The Toronto Star*, April 18, 1993. Reprinted with permission — The Toronto Star Syndicate. **"Back" to the Future** by Portia Jorgenson adapted from "Next Best Buys." Reprinted from *Zoot Capri, The Magazine*, Winter 1990-91 issue. Copyright – The Alberta Alcohol and Drug Abuse Commission.

Illustrations on pages 11, 123, 144, 152, 154: Kim La Fave

THE EDITORS

Christine McClymont was born in Scotland, but came to Canada at an early age. For many years, she has been compiling anthologies for Nelson Canada series such as *Networks*, *In Context*, and *Features*. Christine enjoys hiking and cross-country skiing, and is actively involved with the Toronto Chamber Society.

James Barry is Chairman of the English Department at Brebeuf College School, North York, Ontario. He is the editor of the poetry anthologies *Themes on the Journey*, *Departures*, *Side by Side*, and *Poetry Express*, as well as an annual student writing anthology, *Triple Bronze*. Besides teaching, his special interests are sports (especially hockey), music, and student writing.

Berenice Wood is an instructor in the Faculty of Education at the University of Victoria. She has been an English Department Head and Secondary English Coordinator, and was responsible for developing the current Language Arts–English provincial curriculum for the B.C. Ministry of Education. Her interests include gardening, writing, and all generations of *Star Trek*.